Mythologies of Transhumanism

Michael Hauskeller

Mythologies of Transhumanism

palgrave
macmillan

Michael Hauskeller
Department of Sociology, Philosophy and Anthropology
University of Exeter
Exeter, UK

ISBN 978-3-319-39740-5 ISBN 978-3-319-39741-2 (eBook)
DOI 10.1007/978-3-319-39741-2

Library of Congress Control Number: 2016956560

Printed on acid-free paper

This Palgrave Macmillan imprint is published by Springer Nature
The registered company is Springer International Publishing AG
The registered company address is: Gewerbestrasse 11, 6330 Cham, Switzerland

Acknowledgements

I have so far published two monographs on the philosophy of human enhancement—namely, *Better Humans? Understanding the Enhancement Project* (Acumen/Routledge 2013) and *Sex and the Posthuman Condition* (Palgrave Macmillan 2014). In *Better Humans?* I focused on the concept of human enhancement as an enhancement of the human *as* a human and looked at the normative assumptions that underlie the various proposals for human enhancement in different areas. *Sex and the Posthuman Condition* focused on the role that sex plays in transhumanist and proto-transhumanist visions of our post-human future and what that reveals about the way we understand ourselves in relation to other people. What I have not systematically explored in either of these two books is the peculiar function that the idea of human nature has in the public and academic discourse on the desirability of radical human enhancement. Much of the academic literature on human enhancement engages primarily with the ethics of human enhancement and shows little interest either in the question of what it means to be human (and what it means to ask that question) or in the narrative context of the arguments that are exchanged between proponents and opponents of radical human enhancement. In this book, I attempt to fill that gap.

Some previously published material has been reworked and incorporated. Chapter 2 draws on "Reinventing Cockaigne: Utopian Themes in Transhumanist Thought", *Hastings Center Report* 42/2 (2012): 39–47, and "Utopia", in *Post- and Transhumanism. An Introduction*, eds. Stefan Sorgner & Robert Ranisch, Frankfurt: Peter Lang 2014, 101–108; and Chap. 3 on "Making Sense of What We Are. A Mythological Approach to Human Nature", *Philosophy* 84 (2009), 1–15. Chapter 4 is based on "Messy Bodies, or Why We Love Machines", in *Making Humans*, ed. Alexander Darius Ornella, Freeland: Interdisciplinary Press 2015, 93–106; Chap. 5 on "Human Nature from a Transhumanist Perspective", *Existenz* 8/2 (2013): 64–69, "Pro-Enhancement Essentialism", *AJOB Neuroscience* 2/2 (2011): 45–47, and "Nietzsche, the Overhuman, and the Posthuman", *Journal of Evolution and Technology* 21/1 (2010): 5–8; and Chap. 7 on "A Cure for Humanity? The Transhumanisation of Culture", *Trans-Humanities* 8/3 (2015): 131–147, "Clipping the Angel's Wings: Why the Medicalization of Love Is Still Worrying", *Cambridge Quarterly of Healthcare Ethics* (2015): 361–365, and "Being Good Enough to Prevent the Worst", *Journal of Medical Ethics* 41 (2015): 289–290. A version of Chap. 8 was published as "Levelling the Playing Field. On the Alleged Unfairness of the Genetic Lottery, in *The Ethics of Human Enhancement. Understanding the Debate*, eds. Steven Clarke et al., Oxford: Oxford University Press 2016, 198–210. Chapter 9 makes use of a review of Steve Fuller and Veronika Lipinska's *The Proactionary Imperative*, which was published in *Sociology* (2015, online first). A version of Chap. 10 will appear as "Automatic Sweethearts for Transhumanists", in *Philosophy and Ethics of Sex Robots*, eds. John Danaher & Neil Levy, Harvard: MIT Press.

I am grateful to many people for the exchanges we had on topics related to the book and for inspiring my thinking about them, too many to mention them all. However, I feel especially indebted to Erik Parens, Greg Kaebnick, Brian D. Earp, John Danaher, Kyle McNease, Lantz Fleming Miller, Eliza Kolovou, Alexander Ornella, Inmaculada de Melo-Martin, and Mary Midgley; to my PhD students Lewis Coyne, Alexander Badman-King, Taline Artinian-Papazian, James Watson, and Peter Sjöstedt-H; and last but by no means least, to my wife, fellow philosopher and favourite storyteller, Elena Teodora Manea Hauskeller.

Contents

Prologue: Being in the Way

When people who don't know me ask me what I do, I tell them I'm a philosopher. When they ask me what I specialise in, I tell them that I am mostly, even though I've never been entirely comfortable with that label, an ethicist. This used to be a good thing, or at least not a bad one, but things are changing. People like me are now increasingly being described as, at best, a nuisance, and at worst, a threat to human well-being and possibly even survival.

In this vein, Steven Pinker, the well-known psychologist and bestselling author, has recently (1 August 2015) published an opinion piece in the *Boston Globe*, entitled "The Moral Imperative for Bioethics", in which he chides ethicists for hindering the progress of our species. According to Pinker, biotechnology could do amazing things for us if we only stopped hampering research by raising flimsy ethical concerns about it, which is not helpful at all. Scientific and technological progress is already slow enough as it is, and given the "vast increases in life, health, and flourishing" that biomedical research promises, every day we lose worrying about the ethics of the matter is one day too many. While biotechnological research is urgently needed to rid us of all sorts of terrible diseases, what we most certainly do *not* need are professional worriers who call themselves ethicists, second-guessing every promising new development and thus stalling scientific and technological progress by throwing "nebulous but sweeping principles such as 'dignity', 'sacredness', or 'social justice'" in its way. A *true* ethicist, Pinker decrees, would realize that there is in fact only one valid moral imperative they should promote and follow, namely to "get out of the way".

For Pinker and others like him, ethics is a luxury that we cannot afford. People are dying, people are suffering. The biotech industry is attempting to do something about it, working very hard to succeed, while the "so-called ethicists" are attempting to prevent this from happening. Humanity is painfully pushing a rock up a hill, while all that ethicists are doing is help push it back down again. For Pinker, it is as simple as that. Except, of course, it is not. Surprising as it may be, it is in fact *not* the primary goal of us ethicists to make life difficult for those who want nothing but to make the world a better place. Ethics is not about issuing "red tape, moratoria, or threats of prosecution" (although ethical reflection *may* occasionally give rise to all that). Instead, ethics is about making sure that we know *what* we are doing and *why* we are doing it, that the path we are following is really the path we *want* to be following, and that the place where this path is likely to lead us is really the place where we want to end up being.

We all, naturally, want things to be better than they are, if that is possible. We all want progress. But just as nothing is ever better *as such*, but only ever in certain respects, there is no such thing as progress as such, or in the abstract. We are not sitting in an evolutionary elevator that has only two directions: up and down. Instead, there are many different ways of going up and going forward, many different ways of going down and backwards, and many different ways of going sideways, or around in circles, or of moving without any clear direction at all. Moreover, the ways that lead upwards in *some* way may also lead downwards in some *other* way. Things are usually more complex than we would like, and for this very reason, also more complex than we may care to acknowledge. In order to progress, to step forward, you need to have a goal, or at least have made up your mind about a direction. Spending a thought or two on the reasons for choosing that particular goal or direction before you start running doesn't seem like such a bad idea. And that is all we are doing when we are engaging in ethical reflection. The one question that ethicists keep asking is whether the things we do or propose to do *are actually good for us, all things considered*. Would we really prefer that this question be no longer asked?

It is highly naïve to assume that all biomedical research will necessarily benefit some of us, let alone humanity as a whole. What is powerful enough to save us is also powerful enough to harm us. To demand that such research not be regulated in any way because *some* of it might eventually help us find a cure for Alzheimer's and other diseases is like saying that politicians in government should be granted unlimited legislative, judiciary, and executive power and not be checked in any way because *some* politicians might actually use that power for the good of the people: we just need to trust that they know best and that they want only what is best for us. But why should we believe that?

I suppose most people would agree that granting such unlimited powers to politicians would be a singularly bad idea, and that, even if scientists and biotech firms were generally smarter and more trustworthy than politicians, they are certainly not trustworthy enough that we could afford not putting any regulations and safeguards in place and thereby retaining some measure of control. Ethical reflection helps us determine the nature and extent of those necessary safeguards. Ideas such as human dignity, sacredness, and social justice may strike tough-minded empiricists like Pinker as decidedly airy-fairy and not worth serious consideration. But even though they *are* a bit airy-fairy, for many of us, they do capture something that is both elusive and very real, a sense perhaps that living disease-free and surviving as long as possible is not all that matters, that sometimes more is at stake, that there are other dimensions of our life and experiences that are important to us and for us, whatever they may be. Ethicists are the ones who try to figure out what those dimensions are and why they matter to us. They are not the professional doomsayers that some like to depict them as. Their role is more that of psychopomps who guide us from the present to the future, providing secure passage, making sure that we get there safe and sound.

1

Introduction: From Logos to Mythos

The Saturday edition of the British newspaper *The Guardian* comes with a magazine, the *Guardian Weekend*, which always contains a standardised interview with some public figure. It is called "Questions & Answers". The questions are always the same, except that not all of them are asked each time. Among them are: "What is your greatest fear?", "Who would play you in the film of your life?" and "What is the worst thing anyone's ever said to you?" The answers are often quite telling and can reveal a lot about the person who gives them. But my favourite question is this one: "What would your superpower be?" I am always disappointed when it is not asked.

The answers that people give to that question do not merely tell us something about them, but also about the dreams and fears that are prevalent in our culture. Here is a small selection of the answers that people have given to that question. At the top of the list is "immortality", just one word, without further explanation. The answer betrays a firm belief in one's own significance. Death is seen as a personal affront. Interestingly, immortality is the superpower of choice for the singer Tom Jones, the philosopher Martha Nussbaum, and the founder of *Playboy* magazine, Hugh Hefner.

M. Hauskeller, *Mythologies of Transhumanism*,
DOI 10.1007/978-3-319-39741-2_1

Then there are the do-gooders like the former German tennis champion Boris Becker: "To make the world a better place." The same desire is expressed, in less general terms, by the former British Prime Minister Gordon Brown in his answer: "Magic medicine. I'd love to be able to fix things for the sick and injured. The NHS is the closest thing to it – that's why I'm such a passionate advocate of our system and its doctors and nurses." No wonder he didn't get re-elected. In comparison, the former British leader David Cameron and his erstwhile sidekick Nick Clegg appear refreshingly honest and surprisingly unanimous when it comes to their secret dreams. Cameron: "Teleporting – it would save a lot of travel time." Clegg: "Easy. Teleporting." What an interesting mixture of boyish romanticism (Beam me up, Scotty!), pragmatism, and professionalism, which expresses perfectly both their character and the role they chose to play.

More sinister dreams are voiced by Danny DeVito ("To have people do things the way I want") and Lisa Marie Presley ("I'd be a witch"), and more realistic and modest ones by an ageing Roger Moore ("Being able to get out of a chair without clicking knees or an aching back") and the British actor and political activist Tony Robinson, who, in days of yore, brilliantly portrayed Rowan Atkinson's servant Baldrick in *Blackadder* ("Having to wee only once a day").

Then there are the cultured ones like the conductor Daniel Barenboim ("To travel in time – in order to spend a day with Mozart") and those who are—how shall I put it?—more at home in the flesh like the late singer Amy Winehouse ("Super sexuality").

Very popular is also yet another form of easy locomotion. Bruce Willis: "Flying." Cuba Gooding Jr.: "I dream about flying." And, last but not least, my absolute favourite, with an unbeatable dry irony, Margaret Atwood: "The flying-around thing. With a cape."

These are the things we dream of: having the ability to live without end, to never have to face the annihilation of our self that awaits all human beings, and indeed, all living creatures; living in a world from which all the bad things that plague us have disappeared and everything that is broken can be fixed; gaining full control over space and time so that spatial and temporal distances no longer stand in the way of our desires; having the power to do whatever we want, always, and to make others do what we want them to do; no longer having to put up with our ageing and frail bodies; being able to remove ourselves from the down-dragging weight of our earthbound existence; and, last but not necessarily least, having a more fulfilling and less demanding sex life.

All these dreams echo through our culture, and have been doing so since we began to reflect and tell stories about ourselves and our place in the world, and they all find a natural home today in the philosophy that is known as transhumanism. Transhumanism promotes the use of biotechnologies to modify and improve our nature, to transform us into a different kind of being. Guiding ideas are the desirability of human self-design, the elimination of all suffering and expansion of human autonomy, immortality, and ultimately the complete defeat of (human) nature. Transhumanists believe that we are finally on the brink of making the ancient dream of transcending the human condition come true. However, it has not been sufficiently noticed and appreciated yet that transhumanist arguments in support of radical human enhancement rest on certain narratives about what it means to be human and what a good human life consists in. The discussion usually proceeds on the level of the surface arguments and largely ignores the storytelling (*mythos legein* or mythologies) that surrounds them and on which their plausibility crucially depends. In this book, we are going to explore those *narratives of human nature* that inform the transhumanist discourse because it is only by uncovering the, in that sense, mythological foundations of transhumanism that we can properly assess its plausibility as a philosophy.

I use the term "mythology", which combines the Greek words *mythos* and *logos*, to signify not the learned study of ancient myths, but rather the telling of stories that carry a deep cultural or spiritual significance and provide an explanation and justification for certain practices that are either already established, in the process of being established, or desired to be established. By using this term, I also wish to emphasise the fact that seemingly rational philosophical arguments are rarely, if ever, self-sufficient. The concepts that are employed in those arguments are "embodied in myths and fantasies, in images, ideologies and half-beliefs, in hopes and fears, in shame, pride and vanity. Like the great philosophers of the past who helped to shape our tradition, we need to start by taking notice of these." (Midgley 2011, 128). They are, in other words, embedded in narrative structures of meaning. All philosophies tell a story: about what it means to be human; about what is worth doing, desiring, and fighting for; about good and evil; and, ultimately, what life is all about. As Steven B. Harris has pointed out, myth "is not only religion, of course; it is something more inclusive, broadly encompassing such things as rituals and beliefs. But myth is especially the collection of stories we tell to give ourselves a narrative psychological framework with which to deal with the world. In that larger sense, myth includes (but is not limited to) any story that answers the difficult questions of life: Who am I? Where did I come from? Where am I going? What is the far future going to be like? What is expected of me? Who are the heroes? (What is the Good? What defines Cool?) What is going to happen to me when I die?" (Harris 1996, 45).

What we commonly see as the progression from *mythos* to *logos*, story to argument, emotion and intuition to reason and rational thinking, and subjectivity to objectivity is not, and can never be, complete. Logos always remains firmly rooted in mythos, which gives logos its direction and purpose. In this sense, logos always points back at mythos. Myth, says Neil Gillman, is "an imaginative way of shaping complex data into a structure of meaning. Myths of this kind can neither be verified nor falsified. They can only be challenged by an alternative myth, and they can be testified against." (Gillman 2004).

Some years ago, in summer 2012, I had some time to spare in London and, prompted by a poster that I happened to come across, visited the Superhuman exhibition in the Wellcome Collection. I hadn't heard anything about it, so was curious about what to find. The exhibition was rather small, and the exhibits appeared curiously random. There was a little statue of the flying Icarus (without a cape), obviously meant to be a potent symbol of human enhancement (which struck me as a rather unfortunate choice given the disastrous consequences of Icarus's attempt to fly to freedom); various devices that were thought to represent early enhancements, such as glasses, dildos, a set of teeth, and running shoes; pictures of comic book superheroes; prosthetic limbs; a video of a pretty, naked female artist whose body is being marked by a plastic surgeon; other "artistic" films dealing in some way with the human body, its frailty, and the possibility of its transformation (providing, among other things, a good view between the wide open legs of another naked female, which seemed of particular interest to the group of college students that also happened to be there); a microchip similar to that implanted in the arm of Kevin Warwick; a robot wheelchair that moves by itself through a room; and, last but not least, video clips showing some of the usual enhancement suspects—Julian Savulescu, John Harris, Andy Miah, Anders Sandberg, Barbara Sahakian, and Bennett Foddy—giving speeches about both the desirability and the ordinariness (same old, same old) of human enhancement:

Sandberg: Why should we have to exercise in order to be strong and fit? Couldn't we achieve that using a pill or some other means? (…) Human technology is something natural. (Wellcome Collection 2012, 30)

Miah: During the ancient Olympic Games 2700 years ago (…) athletes were using technologies or natural products to make their products more capable for the sporting environment. Athletes would rub oil on their bodies to protect themselves against the baking heat of Olympia when they do their marathon (…). 2700 years later, not much has changed: athletes are still experimenting with technology to try and push themselves as fast and as hard as they can. (Wellcome Collection 2012, 32)

Foddy: People often think of anti-ageing medicine or lifespan enhancement as somehow being outside the domain of medicine but almost every medical intervention that we've made over the past 150 years has extended human life, and to a lesser degree, human youth. Really it's the project of medicine to extend human life, to enhance life and human youth and to defeat age and death. (Wellcome Collection 2012, 37)

Harris: Enhancement is part of medicine. (…) I'm mystified by the resistance that human enhancement faces. (…) we've got to enhance ourselves; we know we have (…); there is literally no alternative. (Wellcome Collection 2012, 38)

Both the exhibits and the comments by the supposed experts strongly suggested that human enhancement is nothing new, nothing out of the ordinary. In fact, enhancing ourselves is what we have always done. That there might be a difference between external devices such as glasses and the permanent transformation of the human body, between the replacement of lost limbs and functions and the attempt to gain superhuman abilities, is vehemently denied. Yet oddly enough, little of the actual exhibition had much to do with "superhumans". There was no clearly articulated idea of what would turn us into better humans, or into beings that are better than humans, other than perhaps the image of the comic book superhero. The main objective seemed to be to blur distinctions, to make us believe that everything is the same as everything else, that we are all enhanced anyway, so that the envisaged technological enhancement of the human is nothing we have reason to worry about. On the contrary. This message is reinforced by an appeal to human nature, to what being human is all about, in the introduction to the exhibition brochure provided by the exhibition's curator Emily Sargent:

> While human enhancement might initially seem to be the preserve of science fiction, the exhibition examines the subject through the lens of broader human experience. Initial fears that enhancement might compromise our core values are dispelled as we unravel the subject and face the possibility that it is our very desire to improve ourselves that makes us human. The extraordinary range of objects, artworks and ideas that have been brought together for this exhibition reflects this. *Superhuman* highlights the ingenuity displayed in the past to overcome obstacles or conquer new frontiers, while offering a glimpse of what we might look forward to in the future. (Wellcome Collection 2012, 6)

So there is nothing to fear really and everything to look forward to. Unfortunately, the exhibition itself did not dispel any fears at all, simply because it did not even address them. Neither did it show us what we "might look forward to", unless this is looking like a robot or perhaps having prosthetic devices that are indistinguishable from real limbs. And there was certainly nothing there that would suggest, let alone demonstrate, that "it is our very desire to improve ourselves that makes us human". This claim, however, is the myth at the centre of the argument that is being made here, a bit of storytelling that helps us to make sense of it all, to give a particular meaning to the rapid social and cultural changes we experience and envisage. It is a claim that cannot be verified nor falsified. It can only be, as Gillman says, challenged by an alternative myth or testified against.

Here is an alternative myth: four years ago, during the same summer the *Superhuman* exhibition took place, London also hosted, with a huge hullabaloo, the Olympic Games as well as its supposedly ugly little sister, the Paralympics. To promote and kindle interest in the latter, various TV channels broadcast a 90-second video promo, produced by a Channel 4 team, advertising the event. In the video, we see a number of disabled athletes, training hard, faces full of determination, visibly ready to kick ass, and get a glimpse of the events that caused them to be what they are today: terrible accidents, war, genetic defects. Yet the film is made in such a way that those athletes do not come across as disabled at all. On the contrary, they appear to be immensely abled, despite the fact that they have got a limb or two missing or are confined to a wheelchair. There is no weakness to be seen, nothing that is apt to arouse compassion, or worse, pity in us. What we feel instead, and are meant to feel, is admiration, or even more than that: something more akin to awe. This impression is reinforced by four lines of text superimposed over the pictures: "Forget everything you thought you knew about strength./ Forget everything you thought you knew about humans./ It's time to battle./ Meet the superhumans." It was thus suggested that we regard the athletes competing at the Paralympics not as disabled, but on the contrary, as superabled, not as less than human, as deficient in some way, but as more than human. I find this utterly remarkable, because it turns the usual perspective on its head. Superhumans are normally pictured as gifted with special physical or cognitive abilities that allow them to do things that no mere human can do. They can fly or have X-ray eyes or read minds or bend time or are indestructible, or what have you. This is the kind of fantasy that informs much of our current thinking about human enhancement. The radically enhanced human or post-human that transhumanists and others envisage is really not much different from a comic book superhero. Both are able to do things that mere humans cannot do for the simple reason that, for them, the boundaries that determine our human existence no longer exist. They have overcome those boundaries by making them disappear. What the Channel 4 video spot about the Paralympics suggested, though, is that real strength does not show itself in a limitless existence, in the creation of an environment that no longer presents any obstacles to the satisfaction of our will, that is, in other words, in virtual omnipotence. Rather, real strength consists in the spirit. Instead of leaving behind all boundaries, we become more than human by deciding to live with them, but at the same time refusing to let ourselves be bullied by them. We prove both our humanity and superhumanity by refusing to buckle, by putting up a good fight, and by accomplishing great things despite our limitations. Thus, it is not human enhancement, at least not the kind of human enhancement that is commonly discussed as such, that will make us superhuman. If anything, it is resolve, and courage, and related virtues of the mind and heart.

Clearly, this is not a story that we would expect a transhumanist to tell. It is simply not compatible with the transhumanist worldview. So what kind of stories do transhumanists tell? They certainly perpetuate the myth of inevitable progress and the myth of science's "omnicompetence" identified by Mary Midgley (2011, 18). Michael Burdett (2015, 133) defines the myth of progress as "the belief that history/society/humanity has advanced, and is continuing to advance, and will advance in the future". The myth entails that the present is deficient (with respect to the expected future progress). Transhumanism radicalises this view by pairing it with the myth of science's omnicompetence, the belief that science can solve all problems and will ensure our continuing advancement to ever higher levels of existence and experience. "In terms of narrative," the literary theorist Northrop Frye once remarked (1957, 136), "myth is the imitation of actions near or at the conceivable limits of desire (…). The fact that myth operates at the top level of human desire does not mean that it necessarily presents its world as attained or attainable by human beings. (…) The world of mythical imagery is usually represented by the conception of heaven or of Paradise in religion." For the transhumanist, Paradise *is* attainable. Science and technology make it so.

In the following chapters, I will explore these and other myths that I take to be central to the transhumanist enterprise, focusing on *verbally* transmitted stories (or elements in the overarching transhumanist story) rather than *images*, which of course also play an important role in cultural storytelling. Transhumanist imagery featuring human–machine mergences and all sorts of other super- and post-humans are all over the media, functioning, as Frommherz (2013) has argued, as self-replicating cultural memes in the service of propagating transhumanist ideas such as an understanding of evolution as progression and the sublimity of technology, in order to "perpetuate the dream of an autonomous and omnipotent humanity" (Frommherz 2013, 148). Images and words work together to get those ideas a firm hold in our collective mind and to create and promote the transhumanist mythology.

The next chapter will start off our exploration of transhumanist storytelling with a discussion of transhumanism as a form of utopianism. Transhumanist writings are rich with utopian ideas and images that can be traced back to ancient and medieval myths, dreams, and hopes. By analysing the former in the light of the latter, I try to show that the persuasiveness of transhumanist arguments for radical human enhancement crucially depends on their utopian content, and that this seriously undermines transhumanists' self-proclaimed commitment to critical rationality and, consequently, diminishes the weight that we should give to their arguments.

Bibliography

Burdett, Michael S. (2015) "The Religion of Technology: Transhumanism and the Myth of Progress", in *Religion and Transhumanism. The Unknown Future of Human Enhancement*, eds. Calvin Mercer and Tracy J. Trothen, Santa Barbara: Praeger, 131–147.

Frommherz, Gudrun (2013) "Memetics of Transhumanist Imagery", *Visual Anthropology* 26/2: 147–164.

Frye, Northrop (1957) *Anatomy of Criticism*, Princeton, NJ: Princeton University Press.

Gillman, Neil (2004) "A Jewish Theology of Death and the Afterlife", in: *The Fountain of Youth*, ed. Stephen G. Post and Robert H. Binstock, Oxford: Oxford University Press, 94–108.

Harris, Steven B. (1996) "The Immortality Myth and Technology", in: *Immortal Engines. Life Extension and Immortality in Science Fiction and Fantasy*, ed.

G. Slusser, G. Westfahl Westfahl, and E.S. Rabkin, Athens and London: University of Georgia Press 1996, 45–67.
Midgley, Mary (2011) *Myths We Live By*, with a new foreword by the author, London: Routledge Classics.
Wellcome Collection (2012) *Superhuman. Exploring Human Enhancement from 600 BCE to 2050*, London.

2

Anxious Dreams of a Better World

Transhumanism was defined by the evolutionary theorist Julian Huxley (1957, 17) as the belief that the human species can and should transcend itself "by realizing new possibilities" of and for human nature.[1] This belief, which lay dormant for several decades, is currently supported by a growing number of natural scientists and philosophers. Not all of them refer to themselves as transhumanists, but they all advocate the development and use of new technologies that promise to help us overcome familiar biological limitations and become what we allegedly have always wished to be. A radical transformation of human nature is sought and demanded, in the name of reason, science, and progress, and in the spirit of enlightenment and humanism. Transhumanists want to do something against the "terrible fact of death" (More 1996), and advocate social, mental, and physical improvement not only of individuals but also of the whole species, which, they claim, will also make us happier and less prone to suffering. Nick Bostrom, Director of the Future of Humanities Institute at the University of Oxford, declares: "I want to help make the world a better place" and speculates about "lives wonderful beyond imagination" that future, radically enhanced human or "post-human" beings might enjoy.[2] Similarly, John Harris, who is arguably one of the most influen-

M. Hauskeller, *Mythologies of Transhumanism*,
DOI 10.1007/978-3-319-39741-2_2

tial British bioethicists today, claims that "taking control of evolution and our future development to the point, and indeed beyond the point, where we humans will have changed, perhaps into a new and certainly into a better species altogether" is "nothing short of a clear imperative to make the world a better place" (Harris 2007, 4–5). Claims such as these, which are becoming increasingly common, reveal a conspicuous proximity to utopianism.

Utopias can be loosely defined as "man's dreams of a better world" (Eurich 1967, vii), or perhaps a *perfect* world, with perfect human beings or at least human beings that are as perfect as they can be in a perfect (social, political, or technical) environment. Transhumanist visions of our post-human future evoke not only mythical places such as the Land of Cockaigne, the Isles of the Blessed, or the Golden Age, in which men lived like Gods. They also echo the promises of alchemy and later of modern science to secure wealth and happiness for all human beings.

The proposed transition from the human to the post-human via radical enhancement is typically justified by a speculative account of all the fantastic things and experiences that await us (or if not us personally, then at least humanity) once we have achieved post-human status. Bostrom is particularly articulate in describing the many and practically boundless delights of post-humanity: "You have just celebrated your 170th birthday and you feel stronger than ever. Each day is a joy. You have invented entirely new art forms, which exploit the new kinds of cognitive capacities and sensibilities you have developed. You still listen to music – music that is to Mozart what Mozart is to bad Muzak. You are communicating with your contemporaries using a language that has grown out of English over the past century and that has a vocabulary and expressive power that enables you to share and discuss thoughts and feelings that unaugmented humans could not even think or experience", and so on and so forth (Bostrom 2009, 112). Basically, everything will be much, much better (and easier). In his *Letter from Utopia*, in which one of those fortunate post-humans of the future addresses us merely humans, we are reminded of those few and all-too-short precious moments in which we experience life at its best, only to be told that those moments are nothing compared to the bliss permanently experienced by the post-human: "And yet, what you had in your best moment is not close to what I have now – a beckoning scintilla at most. If the distance between base and apex for you is eight kilometres, then to reach my dwelling requires a million light-year ascent. The altitude is outside moon and planets and all the stars your eyes can see. Beyond dreams. Beyond imagination." Post-humans will no longer be cursed with ageing bodies, and will no longer have to die; they will know and understand things that are entirely beyond our reach now; and above all, they will have lots and lots of pleasurable experiences: "Pleasure! A few grains of this magic ingredient are dearer than a king's treasure, and we have it aplenty here in Utopia. It pervades into everything we do and everything we experience. We sprinkle it in our tea" (Bostrom 2010, 5). The letter ends with an urgent call to bring the post-human into existence and is signed by "your possible future self".

There is nothing very unusual about the utopian outlook that Bostrom (until recently)[3] endorses so unabashedly. On the contrary, it is rather common and apparently shared by many who see humanity's salvation in emerging and converging technologies and technological growth in general. The scientists and US government officials who authored the 2002 landmark report "Converging Technologies for Improving Human Performance", commissioned by the US National Science Foundation and the Department of Commerce, seriously expected that through the convergence of nanotechnology, biotechnology, information technology, and cognitive science, we would soon be able to solve all the world's problems. Technological progress would result in "world peace" and "evolution to a higher level of compassion and accomplishment" (Roco and Bainbridge 2003, 6). More importantly, it would also lead to "a golden age of prosperity" (291) and "economic wealth on a scale hitherto unimaginable" (293). Economic wealth is here clearly seen as both necessary and sufficient for permanent human happiness, where the latter, in well-tried utilitarian fashion, is equated with unlimited access to, and enjoyment of, pleasurable experiences. This essentially materialistic and hedonistic understanding of human progress is reminiscent of the medieval legend of the Land of Cockaigne, where supposedly "no one suffers shortages/ the walls are made of sausages" and "lovely women and girls may be taken to bed/ without the encumbrance of having to wed" (Pleij 2001, 33, 39). Transhumanists occasionally betray similar sentiments and ideals.

David Pearce, for instance, who in 1998, with Nick Bostrom and a few others, drafted the "Transhumanist Declaration" (the founding document of the World Transhumanist Association) and who favours a negative utilitarianism whose aim is the abolition of all suffering, is equally optimistic (and equally hedonistic in his outlook): "Over the next thousand years or so", he knows, "the biological substrates of suffering will be eradicated completely", and consequently, the "states of mind of our descendants (…) will share at least one common feature: a sublime and all-pervasive happiness" (Pearce 1995, 0.1). It will be nothing less than a "naturalisation of heaven", where we "will have the chance to enjoy modes of experience we primitives cruelly lack. For on offer are sights more majestically beautiful, music more deeply soul-stirring, sex more exquisitely erotic, mystical epiphanies more awe-inspiring, and love more profoundly intense than anything we can now properly comprehend" (0.4.). "As an exercise, the reader may care briefly to summon up the most delightful fantasy (s)he can personally conceive. Agreeable as this may be, states of divine happiness orders of magnitude more beautiful than anything the contemporary mind can access will pervade the very fabric of reality in generations to come. Even the most virile of imaginations can apprehend in only the barest and formal sense the ravishing splendour that lies ahead" (1.7.).[4]

In the same vein, to add just one more example, Gregory S. Paul and Earl D. Cox, in their celebration of cyberevolution, ask us to consider "the advantages of being able to learn and understand anything your mind desires in a few minutes. Imagine yourself a virtual living being with senses, emotions, and a consciousness that makes our current human form seem a dim state of antiquated existence. Of being free, always free, of physical pain, able to repair any damage and with a downloaded mind that never dies" (Paul and Cox 1996, xv).

The exuberant rhetoric that marks all those descriptions frames the more serious and, as I am happy to admit, occasionally rather sophisticated philosophical arguments that Bostrom and some of his more academically inclined fellow transhumanists such as John Harris or James Hughes have presented. The rhetorical framing, however, is far more than mere decoration and literary flourish, for the arguments are all based on the presumption that the proposed changes of the human condition will have immensely desirable effects. The rhetoric disguises the fact that we actually know very little about what it would be like to be post-human and that we cannot be certain that the world we are going to create by taking the path of radical enhancement is anything like the world described so imaginatively by its ardent proponents. In fact, the whole idea of being able to fulfil all our desires and to live a life of pure joy that allegedly lies ahead of us betrays clearly enough its mythological roots. Just as the medieval Land of Cockaigne, in which food and drink rains down from the sky, sexual restraints no longer exist, and nobody has to die or fall ill, or the Fountain of Youth, which occurs in one form or another in countless legends and which has the power to return to us what we often miss more than anything else—namely, our lost youth, the transhumanist account of post-human existence is quite obviously a wish-fulfilment fantasy. And there is not even much difference between the themes that inform the mythological fantasies, on the one hand, and the transhumanist fantasies, on the other. Sensual pleasures are still very important (with sexual pleasures ranking particularly high), and so is youth and youthful vigour, and perhaps more than anything else, the freedom to do as one pleases and not to be restricted in any way. As the legend of Cockaigne has it: "Lovely women and girls may be taken to bed,/Without the encumbrance of having to wed./ Nothing sinful about it, no one feels shame,/For their custom in this is not to lay blame" (Pleij 2001, 39). If we compare this, for example, to de Val's and Sorgner's *Metahumanist Manifesto* (2010), we find basically the same idea of boundless sexual liberty, although somewhat intellectualised by pseudo-radical jargon: "Metasex not only challenges the dictatorship of anatomical, genital and binary sex, but also the limits of the species and intimacy. Pansexuality, public sex, poliamoria, or voluntary sexwork are means to redefine sexual norms into open fields of relationality, where modalities of affect reconfigure the limits of kinship, family and the community."

Yet whatever the details of the dreamworld that individual transhumanists conjure up, they usually agree that the radically enhanced post-humans will live like the first race of humans did during the Golden Age that Hesiod tells us about, when there was no hard work or grief and no miserable old age: "They had everything good. The land bore them fruit and all of its own, and plenty of it too. (…) And sure when Earth covered over that generation, they turned into holy spirits" (1993, 138–43). The latter we now hope to achieve, after having enjoyed a long life as physical organisms and an accompanying array of bodily pleasures, by uploading our minds to a computer. Hesiod's holy spirits have adapted to modern times and have reappeared as Kurzweil's spiritual machines (Kurzweil 1999), but they are essentially the same.

Transhumanists like Bostrom, however, often put more or at least equal emphasis on what they think of as cognitive enhancement, which plays rather a minor role in ancient myths. Cognitive enhancement is meant to greatly facilitate understanding and the acquisition of knowledge. It is assumed that of understanding and knowledge, one can never have enough. They are regarded as both intrinsically and instrumentally valuable. Intrinsically, because through knowledge and understanding, we get access to intellectual pleasures of which we would otherwise remain ignorant, and instrumentally, because the more we know and understand, the better will we able to control our environment, and the better we will be able to do that, the closer we get to the kind of naturalised heaven that not only Pearce dreams of. Yet by including knowledge and understanding among the prizes that a radically enhanced human can be certain to attain and by connecting them with the mythological themes of effortless pleasure and eternal life, transhumanists revive yet another set of medieval ideas that we can find at work in the beliefs and endeavours of the alchemists.

Contrary to a popular misconception, alchemists were not primarily concerned with the transmutation of base metal into a far more valuable substance, such as silver or gold, but more generally with the overcoming of the boundaries that separated the various kinds of things from each other and that prevented ascendency to perfection. In their desire to know how diverse substances could be transmuted into another, they made discoveries that helped advance the budding sciences of metallurgy, chemistry, and medicine. The Philosopher's Stone, which was thought to make all transmutations possible, was also the Elixir of Life, or was meant to be used to the same effect. The search for material perfection went hand in hand with the search for spiritual perfection, that is, wisdom (knowledge and understanding). It was generally believed that the ability to turn common metal into gold (a substance that neither rusts nor decays) would go along with unlimited wealth, wisdom, and immortality.

"The philosopher's stone is a symbol for the permanence and perfection which man has always sought and never found. The alchemical dream of transmuting base metal into gold was more than a scheme to get rich quick; it was a dream in which death could play no part" (Coudert 1980, 194). This dream was based on a quasi-Aristotelian understanding of nature, according to which everything that is, strives, by its very own nature, for perfection (or, as per Aristotle, for that state of being that represents its generic optimum). Just as the acorn strives to grow into an oak and the body of a child into the body of a woman or a man, common metal likewise yearns to become something better. To turn it into gold is thus not a violation of its nature, but a way of supporting it. It is not as if by doing so we would force things to do or be what they are not meant to do or be, but rather that we simply help them along, assist them in their journey to perfection and thus in the realisation of their own true nature. The same holds true for the expected transformation of the human from a mortal into an immortal being. It is simply a matter of learning to be the kind of being that we have always meant to be.

Transhumanists often show a similar adherence to a crypto-Aristotelian teleology of the human and they are just as happy to exploit its utopian potential. Max More claims that "to halt our burgeoning move forward, upward, outward, would be a betrayal of the dynamic inherent in life and consciousness. We must progress on to transhumanity and beyond". Nick Bostrom (2009) makes extensive use of the metaphor of growing up and tells us that humans are like children who naturally (although with a little help from enhancement technologies) evolve into post-human adults. If we let this happen (and it is a matter of letting it happen rather than actively bringing it about), we will, according to Bostrom, "truly grow up and experience life as it should have been all along".[5] Finally, Gregory Stock, a prolific advocate of biotechnology and human self-design, maintains that human nature is essentially Promethean, so we will, following our own natural ends, progress further into post-humanity whether we like it or not (Stock 2003, 2). This kind of techno-optimism, and indeed techno-determinism, is not uncommon among scientists working in the field today. Biotechnology promises to be the real Philosopher's Stone, that elusive device which the alchemists so desperately tried to find and which would finally give them the power to reinvent the world so that it would match their desires. The modern utopia rests on a strong belief in the transformative and salutary power of science and technology that has always accompanied their rise and been at the root of their success.

While classical social utopias such as Thomas More's *Utopia* or Tommaso Campanella's *City of the Sun* did not seek to radically transform the human condition, this changed with the publication of Francis Bacon's *Nova Atlantis* in 1627, which marked the transition from the traditional social and political utopias to the modern techno-utopia. Bacon envisaged a scientific and technical utopia in which the "Enlarging of the bounds of Humane Empire, to the Effecting of all Things possible" is declared to be the ultimate goal of the fictional society that the novel's narrator encounters on the remote (fictional) island Bensalem (Bacon 1924, 35). One of the main areas of research those islanders engage in seems to be human enhancement. Among the discoveries they have already made by devoting themselves completely to their goal, Bacon lists the production of food that makes "the very Flesh of Mens Bodies, sensibly, more Hard and Tough; and their Strength far greater, then otherwise it would be" (40); the resuscitation of (seemingly) dead bodies, the ability to make animals grow larger or smaller, more fruitful or barren, to change their colour, shape or behaviour; the creation of chimeras (i.e. mixtures of different kinds of animals); and finally, the creation of "Perfect Creatures" (39). Although these experiments are being undertaken with animals rather than humans, the sole reason for conducting them is in order to "take light, what may be wrought upon the Body of Man" (38).

This tradition was continued into the twentieth century by H.G. Wells, who also distinguished the "modern" utopia by its inherent commitment to constant progress: "[T]he Modern Utopia must be not static but kinetic, must shape not as a permanent state but as a hopeful stage leading to a long ascent of stages" (Wells 1967, 5). Transhumanists tend to share this assessment and emphasise the fundamental unboundedness of the enhancement process, which again links them to earlier modes of thought. Belief in human perfectibility has in fact replaced visions of (realistically unattainable) perfection since the late eighteenth century. In 1795, not yet sobered by the fate of the French Revolution, the Marquis de Condorcet declared human beings to be indefinitely perfectible within the boundaries of human nature, and wondered how much more we could expect for the improvement of our cognitive and physical abilities, the extension of our life span, and ultimately, the conquest of death if we only found a way to improve this nature itself, finally "released from the empire of fate and from that of the enemies of its progress, advancing with a firm and sure step along the path of truth, virtue and happiness" (Condorcet 1955, 199–102). Like many transhumanists today, Condorcet was convinced that it was our human destiny to make this step. At the same time in England, William Godwin preached human perfectibility and saw us becoming increasingly godlike, perhaps immortal, with a necessity that is rooted in our nature, for we are all, essentially, godlike beings (Godwin 1971). A few years later, Johann Gottlieb Fichte in Germany predicted that we will eventually gain a complete understanding of nature and that work will cease to be a burden because men will have learned how to reduce mechanical toil. He was also convinced that God will some day reveal a glorious end to humanity, which "is wholly incomprehensible to me here" (Fichte 1908, 157).

All these utopian ideas can be found again in one form or another in transhumanist writings. Transhumanism closely follows the utopian tradition that has developed from its mythological beginnings via the proto-scientific aspirations of alchemy to modern science and the hopes that its steady progress has kindled and nourished. If anything, the utopian element has, in the wake of the biotechnological revolution, become even more pronounced and expressive of further-reaching ambitions. According to Gregory Stock, we will soon "seize control of our evolutionary future" (Stock 2003, 2). Biological enhancement will lead us to "unexplored realms, eventually challenging our basic ideas about what it means to be human", which he thinks is not something to deplore, but rather to celebrate. The "beginning of human self-design" (3) is a good thing without qualification because it promises, for the first time in history, *complete autonomy*. We have come to regard our own physical bodies as external restraints (rather than as the internal condition of being someone at all and thus the source of all the freedoms that we *have* got). They seem to prevent us from being entirely autonomous. For that reason, we need to overcome not only the nature that surrounds us, but also the nature that we are ourselves. Elsewhere, I have called this vision of *complete control* that pervades transhumanist writings the *ultimate utopia*

(Hauskeller 2007). It is present in the unconcealed desire for personal immortality and the acquisition of godlike qualities such as omnipotence, omniscience, and even omnibenevolence. The idea, publicised by Kurzweil and others, that we will one day be able to upload our minds to computers (and thus achieve immortality) is clearly inspired by the same hopes and desires.

By and large, all transhumanists are optimists regarding the future of humanity (Berthoud 2007, 295). They look forward to what lies ahead of us, and embrace without much hesitation the technologies that are supposed to lead us there. They tend to believe that everything will be for the best, and that the best is what we will get if we are only courageous enough to wholeheartedly commit ourselves to scientific and technological progress. Transhumanists do not doubt that humans are special, that reason sets us apart from the rest of nature, and that we all carry the potential in us to ascend the heavens and to be (or live) like Gods—very much in accordance with the very modern human self-understanding that Pico della Mirandola laid down in his *Oration on the Dignity of Man* (1486), which can be seen as the foundation charter of Renaissance humanism. For Pico (1985, 4), humans were, by nature, free to invent themselves, and not confined by any natural boundaries: "Thou art the molder and maker of thyself, thou mayest sculpt thyself into whatever shape thou dost prefer." As humans, we are naturally disposed to change and to progress to higher spheres. It is in our very essence to transgress boundaries, to go ever further on our way to perfection and godliness. This belief is also at the core of transhumanism. Scratch a transhumanist and you will find a humanist underneath.

In this respect, transhumanism is very different from what is sometimes referred to as "critical", "cultural", or "radical" post-humanism. Despite being rather a diverse lot, critical post-humanists are normally decidedly anti-humanist—or "post-human*ist*" (Wolfe 2010, xv)—and hence also deeply suspicious of transhumanist aspirations to create better, even more glorious humans by means of technology. Critical post-humanists generally refuse to see humans as a "superior species in the natural order" (Miah 2008, 72), ontologically distinct from animals, on the one hand, and machines, on the other. They insist that the boundaries between the human and the non-human are rather fluent and, in fact, have always been so: it is just that this fact has become more pronounced and thus more obvious through recent technological advances. This makes the post-human that critical post-humanists talk about an altogether different entity from the post-human of the transhumanists. In contrast to the latter, the post-humanist post-human is not an entity of an imagined future, but an entity that already exists. For the post-humanist, we are already post-human (Hayles 1999) and, in a certain sense, have always been so. The human (as something essentially different from other entities) has never existed. As Halberstam and Livingston, in their seminal collection of articles on "post-human bodies", echoing Donna Haraway's "we are cyborgs" (1985, 191), programmatically declare: "You're not human until you're post-human. You were never human." (Halberstam and Livingston 1995, 8). Thus, "the human" is merely an ideological construct, a myth, and, ultimately, a lie because the phrase suggests that there is an essential distinction between the human and the non-human, while in fact there is not. Any appearance of an ontological difference between humans and machines, on the one hand, and humans and animals, on the other, is merely a discursive practice that "functions to domesticate and hierarchize difference within the human (whether according to race, class, gender) and to absolutize difference between the human and the nonhuman" (Halberstam and Livingston 1995, 10). It is the ideology of the human that critical post-humanists seek to uncover and to attack. The political goal is to rupture and exceed traditional cultural "narratives" of the human and to "destabilize the ontological hygiene of Western modernity" (Graham 2002, 16) in order to overcome historic divisions between class, race, and gender. For this reason, critical post-humanists are equally opposed to so-called "bioconservative" critics of radical human enhancement, such as Francis Fukuyama, Michael Sandel, or Leon Kass, and to transhumanist enhancement enthusiasts. From a post-humanist perspective, both parties commit the same basic mistake: that, although they may have different ideas about what it means to be human, they both believe in the existence of the human, and in the value of being one. Transhumanists welcome and endorse the new technologies because they seem to provide new, far-ranging possibilities for human progress. Post-humanists often do the same, but for other reasons. The increasing incorporation of modern technology into our lives and bodies is a fact that we have to deal with, and whether we like it or not, it is to be welcomed to the extent that it confuses boundaries (e.g. between human and non-human, male and female, physical and non-physical) and forces (or at least allows) us to review and revise the way we are used to look at the world. "The dichotomies between mind and body, animal and human, organism and machine, public and private, nature and culture, men and women, primitive and civilized are all in question ideologically." (Haraway 1985, 205).

There is, of course, a utopian dimension to the post-humanist critique of humanist and transhumanist progressivism and utopianism, which was initially acknowledged by Donna Haraway in her early *Manifesto* (1985, 193): "This chapter (…) is an effort to contribute to socialist-feminist culture and theory in a post-modernist, nonnaturalist mode and in the utopian tradition of imagining a world without gender". Twenty years later, however, she expressed discomfort with her own utopian interpretation of post-humanism. In an interview with Nicholas Gane (Gane and Haraway 2006, 137), she revokes her earlier remark: "It's not a utopian dream but an on-the-ground working project. I have trouble with the way people go for a utopian post-gender world." Clearly, part of Haraway's discomfort with being seen as trying to launch some kind of utopian project stems from her distaste for the goals of transhumanism: "I can't believe the blissed-out techno-idiocy of people who talk about downloading human consciousness onto a chip" (ibid., 146). Yet she still acknowledges the importance of utopian thinking for the purpose of critiquing (and possibly changing) established practices (ibid., 152): "I suppose there is a kind of fantastic hope that runs through a manifesto. There's some kind of without warrant insistence that the fantasy of an elsewhere is not escapism but it's a powerful tool."

It is obvious that Haraway does not share the enthusiasm that most transhumanists seem to feel for the ongoing technification of the life world—she even admits that it is something of a "nightmare" (Gane and Haraway 2006, 150). Likewise, Katherine Hayles, in her influential book *How We Became Post-human* (1999, 1), speaks of the "nightmare" of a downloaded consciousness and contrasts it with a "dream" of her own: "If my nightmare is a culture inhabited by post-humans who regard their bodies as fashion accessories rather than the ground of being, my dream is a version of the post-human that embraces the possibilities of information technologies without being seduced by fantasies of unlimited power and disembodied immortality, that recognizes and celebrates finitude as a condition of human being" (ibid., 5). Other post-humanists express a similarly ambivalent attitude. Thus, David Wills (2008), while embracing what he calls the "technological turn" and claiming that the "human thing" has never been "simply human", but is and has always been in its very essence a "technological thing" (ibid., 3), argues that, precisely because we always already have technology in our backs, we can and should resist a "technology that defines itself as straight-forward, as straight and forward, straight-ahead linear advance, the totally concentrated confidence and pure technological fiat of an unwavering lift-off" and "reserve the right to *hold back*, not to presume that every technology is an advance" (ibid., 6). According to Wills, control and mastery are an illusion, never to be fully accomplished because technology has us, rather than the other way around. In the same vein, though not always for the same reasons, other critical post-humanists such as Elaine Graham also scorn what they see as transhumanists' "technocratic futurism" (Graham 2002, 155) and "libertarian philosophy" (ibid., 159).

However, despite the widespread post-humanist opposition to transhumanist techno-utopianism, the desired and recommended dissolution of all confining boundaries is clearly itself a utopian idea, whether those boundaries are conceived as physical boundaries (as in transhumanism) or rather conceptual, that is, social and political boundaries (as in post-humanism). At the heart of post-humanism is clearly a liberationist ideal: the hoped-for redistribution of difference and identity is ultimately a redistribution of power. Haraway and those who have been following in her footsteps urge us to see the confusion of boundaries that our use of modern technologies forces upon us, not as a threat, but rather as an opportunity to develop resistance to domination: "[C]ertain dualisms have been persistent in Western traditions; they have all been systemic to the logics and practices of domination of women, people of color, nature, workers, animals – in short, domination of all constituted as others, whose task is to mirror the self" (Haraway 1985, 219). Instead of bemoaning the increasing technification of our life world and resigning ourselves to the role of victims, we are asked to use it in order to undermine existing structures of domination. Again, we are told to be brave in the face of new developments and to see them as an opportunity rather than a threat. However, while transhumanists tell us not to be afraid of letting go of the familiar but defective human and paving the way for the unfamiliar but vastly improved post-human, post-humanists ask us not to be afraid of "permanently partial identities and contradictory standpoints" and to suppress and firmly reject the perhaps all-too-human desire for clear demarcations (Haraway 1985, 194). This requires an appreciation of disorder and illogic, and a repudiation of (normative conceptualisations of) health, purity, and stability (Halberstam and Livingston 1995, 13). Katherine Hayles makes it clear that "the post-human" is just as much a construct as "the human". It is not a real entity that is meant to replace the human at some point in the future, but rather a certain point of view, a new way of looking at things and at ourselves: "Whether or not interventions have been made on the body, new models of subjectivity emerging from such fields as cognitive science and artificial life imply that even a biologically unaltered *Homo sapiens* counts as posthuman" (Hayles 1999, 4). Whether we are human or post-human thus entirely depends on our own self-understanding: "People become post-human because they think they are post-human" (ibid., 6). Along the same lines, Elaine Graham (2002) analyses the different "representational practices" that create the differing worlds of the human and the post-human. Technology changes things, but the really important changes, according to post-humanists, are ultimately in the head. Haraway's "Cyborg" was a metaphor for a changed, or changing, perspective. And so is "the post-human" for many cultural theorists. For transhumanists, on the other hand, the post-human is the radically enhanced, virtually omnipotent human of the future.

Haraway famously concluded her *Manifesto* with the statement that she would rather be a cyborg than a goddess. These two words stand for alternative utopias. What distinguishes critical post-humanists from transhumanists is this: while post-humanists would rather be cyborgs than goddesses or gods, transhumanists wish to be both, but if they had to choose, they would much rather be gods.

Now what is the relevance of all this? Would it matter for our assessment of transhumanism if it were indeed a kind of utopianism? If it were based in a strong belief that a kind of earthly paradise awaits us if only we are courageous enough not to shy away from our human destiny and to wholeheartedly embrace all technological progress? Before we can try to answer this question, we need to determine first what exactly the function of utopian ideas and images in transhumanist writings is. It is fairly clear that they provide considerable motivation for the development and endorsement of radical enhancement technologies, and hence very likely that without a prominent display of such utopian fantasies there would be far less willingness to fund research into, and development of, enhancement technologies. Those ideas thus function as a call to arms to prospective followers and investors. In some cases, this is quite obvious, when, for instance, Aubrey de Grey, after promising us an "endless summer of literally perpetual youth" (de Grey and Rae 2007, 335), calls out a "crusade against aging" (x) and expressly asks his readers to "lobby for more funding for rejuvenation research, and for the crucial lifting of restrictions on federal funding to embryonic stem cell research in the United States, by writing letters to your political representatives, demanding change", or better even, to directly donate to the Methuselah Foundation that de Grey founded in 2003 (336). And if we accept his co-author and research assistant Michael Rae's description of de Grey as "tirelessly and courageously bearing Promethean fire to a world yet shivering under the winter of age-related death and decay" (unpaginated dedication), which not accidentally evokes an image that contrasts starkly with the utopian counter-image of the predicted endless summer of perpetual youth, then we may well feel inclined to join the crusade and empty our pockets for the cause. It seems that the brighter the post-human future appears to us, and the bleaker the human present, the more reason we have to abandon humanity and seek to bring about post-humanity. The purported brightness of the future and corresponding bleakness of the present reinforce each other. For this reason, utopian descriptions of the post-human condition are generally complemented with dystopian descriptions of the merely human, for instance, when death is depicted as "the greatest evil" (More 1996) or more imaginatively and memorably as an all-devouring dragon whose "red eyes glowed with hate" and whose "terrible jaws flowed an incessant stream of evil smelling yellowish green slime" (Bostrom 2005, 273). And once the dragon is dead, of course, the future is suddenly wide open for the creation of a "better world" (276).

Transhumanist descriptions of how our post-human future is going to be like are descriptions only on the surface. In fact, their purpose is very different from that of mere descriptions. Their aim is not to describe facts or to express a belief about the (future) state of the world. Instead, they belong to a class of speech acts that John Searle has described as being "not in the business of trying to tell us how things are in the world", but rather of "trying to change the world to match the content of the speech act" (Searle 2010, 11). As paradigmatic examples of such speech acts, Searle mentions promises and commands. Utopian accounts of our post-human future have something of both. They promise us a far better future, a future that is presented as definitely worth pursuing and, in fact, more desirable than anything else. The act of promising entails a commitment, but the fulfilment of the promise may nonetheless presuppose that certain conditions are met. I can promise you that something will occur *if* you do what I tell you to do. My commitment is to the certainty of the outcome, *provided* you collaborate. Post-human utopias are similar to such conditional promises. They are presented as certain outcomes (conveying that 'this fabulous alluring future actually *will* occur!'), but at the same time, as dependent on our willingness to help bring it about and not to throw any unnecessary obstacles in its way. The promise thus borders on a command. We are told to support radical enhancement ('Do this!') and, so that we have a reason to obey, we are promised a hefty reward (an indefinitely extended life span, pleasures beyond anything we can currently imagine, vastly superior understanding, autonomy and complete control, you name it).

This precarious combination of promise and command that underlies the usual accounts of our post-human future makes its utopian character even more significant for a critical assessment of the transhumanist agenda. Utopian ideas and images do not merely serve as motivational aids to get people to support the radical enhancement agenda. They also affect the very arguments that are proposed in favour of human self-transformation and, in particular, in support of the claim that it is our moral duty to develop and use technologies that make this happen. As philosophical arguments, they appear to be self-contained, but in truth, utopian ideas form the fertile soil from which those arguments grow, so without them they would wither and die.

So how relevant is all this for our assessment of transhumanism as a philosophical movement whose explicit goal it is to change the world by changing human nature? It may seem that nothing has been said so far that would necessarily discredit the transhumanist enterprise as such. Why, after all, should it be wrong to dream of a better world and to encourage everyone to help bring it about? Perhaps the alleged similarity between age-old utopian ideas and the transhumanist agenda is merely superficial anyway, and even if it is not, it is far from obvious that there is anything wrong with utopianism as such. On the contrary, it seems that a decent dose of utopianism is the engine of all progress, not only the progress that is yet to come, but also all the progress that we have made in the past to get us to where we are now. Surely we would not want to relinquish that.

Now first of all, the similarities are anything but superficial. What drives transhumanists and their persistent call for radical human enhancement is the same old desire that expresses itself in ancient myths and modern utopias: the desire for overcoming natural restraints, for a life not limited by things that we cannot control. What has changed is merely that, for the first time in history, mainly due to the rapid development of the biosciences and related technologies, it actually seems possible that we will very soon achieve all this: that we will be free of sickness and disease, free of the necessity to die, know everything there is to know, enjoy pleasures without restraint or remorse, and live in complete harmony with others and with ourselves. But even before the science and technology existed that today promises to make all this happen very, very soon—according to de Grey, the first person to live to 1000 years is probably already in his 60s, and Kurzweil (2005, 135–6) expects his Singularity, which represents a "profound and disruptive transformation in human capability", to occur in 2045—people already envisioned such a technology. Bacon's almost 400-year-old description of the things that the scientific community on his fictional island Bensalem have already accomplished bears a remarkable similarity to the things that we are doing, or trying to do, today. This suggests that it is not recent scientific developments that first gave rise to the ambitions for the radical transformation of the human condition that transhumanists encourage us to share, but on the contrary, that it is those ambitions that have kept us looking for the means to complete them until eventually we seem to have found them, or at least to have come very close to finding them. Scientific and technological developments are ultimately driven by non-scientific purposes. All that science can ever do is provide the means to ends whose origin lies beyond science (although it might fuel the pursuit of those ends).

Of course, the availability of the means does make a huge difference in practical terms. Plato's vision of a radically transformed society ruled by philosophers could not do much harm because he lacked the political power to make his vision real. Yet the history of the twentieth century has shown how utopian ideas of an ideal advanced society, when endorsed by powerful rulers determined to change the world according to their vision, for instance, in Hitler's Germany, Lenin's and Stalin's Russia, or Mao's China, tend to be a sure recipe for disaster. Attempts to play out dreams of a better world have usually led to the suffering and death of millions of people. The difference between those dreams and the dreams expressed by transhumanists is marginal. Again, it consists mainly in the means that are expected to bring the transition about. While the political utopians of the early-to-mid twentieth century set their hopes mostly on the reorganisation of economic and social life, transhumanists hope to achieve their goals by means of the rapidly progressing natural sciences and the accompanying advances in technology. But the goals are very similar: "Man will make it his purpose to master his own feelings, to raise his instincts to the height of consciousness, to make them transparent, to extend the wires of his will into hidden recesses, and thereby to raise himself to a new plane, to create a higher social biologic type, or, if you please, a superman." Nobody would be surprised to find this passage in any of the countless transhumanist manifestos. In fact, however, it was taken from a book called *Literature and Revolution* that was published in 1924. The author was Leon Trotsky (2005, 207). Mere coincidence? I think not.

However, I am not suggesting that it is necessarily wrong or bad to dream of a better world or to try to make the world a better place. Of course, important developments have often been driven by utopian dreams of a better world. This better world is always one that allows people's lives to be, in some important (albeit varying) respect, better than they normally are at the time when, and the place where, the dream is dreamt. That imagined world, which compares favourably with the here and now, can be in the past (the Golden Age, Paradise Lost), in the future (Heaven and a New Earth, Paradise Regained, and secularised versions thereof), and even in the present (mythical places like the Isles of the Blessed or

Avalon, but also real places such as America—the "land of the blessed"—for European emigrants in the early 1900s or Communist Russia in the 1930s). So clearly, those dreams fulfil an important function. They serve as a reminder that the world does not *have* to be as it is: that there are other possible worlds that we could live in—worlds in which nobody is poor and where everyone has enough to eat; worlds in which people are not being oppressed and each can say what they please, where everyone counts for one and no one for more than one; worlds perhaps where we do not have to work so hard and where there is more enjoyment, where being alive is an unimpaired pleasure, where there is no suffering, disease, or death, where we are powerful and no longer have to fear anything or anyone. Utopian dreams like these have no doubt stimulated social, scientific, and technological progress. However, we must not forget that they have also led to humanitarian disaster when concerted attempts to make the dream come true failed miserably. Unfortunately, some worlds turn out to be less desirable than they appeared to be in our dreams, and some dreams get compromised by the means thought necessary to realise them. Others are repugnant in their own right, like the utopian dream of a world in which, say, the Aryan race reigns supreme. Clearly not all dreams are worth dreaming, and not all survive their implementation into the real world undamaged. The challenge is to know in advance what will happen if we endeavour to turn utopia into reality.

The problem with the transhumanist dream is that its realisation requires a *radical* transformation of the human condition, and radical transformations, and even all *attempts* at radical transformation, are typically fraught with dangers and uncertainties, as even Nick Bostrom has now begun to realise (Bostrom 2014). This is the reason why we cannot ignore the utopian elements in transhumanist arguments for radical enhancement. They are highly relevant because they effectively conceal the fact that we actually have no idea whether or not the suggested transformations of the human body and mind will really work out the way it is suggested. Yet by dwelling on the glorious future that allegedly awaits us, transhumanists make the risks of such an enterprise appear negligible or at least acceptable, which not only is intellectually dishonest, but also impedes a fair and rational assessment of the actual desirability of radical human enhancement. One striking example of this strategy at work can be found in Nicholas Agar's earlier book *Liberal Eugenics*, where he briefly discusses the risks that radical enhancement may pose and then quickly comes to the conclusion that the precautionary principle can be safely disregarded because enhancement technologies "actually do present potential benefits of a magnitude comparable with the nearly infinite potential penalties imagined by opponents" (Agar 2004, 163).[6] In other words, our enhanced future existence is going to be so good that it is worth taking any risk at all to get there. For the argument to work, the benefits of radical enhancement must be pictured as so immense that a radical transformation of our very nature can plausibly be regarded as worth attempting. And the greater the benefits that are being promised to us, the more likely it may seem to us that at least *part* of what is being promised will actually come true. The need to greatly exaggerate the expected (or promised) gains becomes even more pressing when, as some philosophers do, radical enhancement technologies are described as something that we actually have a moral *duty* to develop, promote, and apply (e.g. Harris 2007, passim). The postulation of a moral duty reaffirms the command character of the utopian description. We are told to act in a particular way because: (a) it will help bring about a "better world" (which mostly means one that perfectly responds to all our desires), and (b) we have a moral obligation to bring it about, precisely because it is a *better* world. In other words, our post-human future is going to be so good that it would be not only foolish to relinquish it, but moreover, a *crime against humanity*.

Second, by reconnecting with the crypto-Aristotelian faith in the "vocation of man" (Fichte), which was prevalent in the late eighteenth century, but goes back at least to the Renaissance and, in particular, Pico della Mirandola's *Oration on the Dignity of Man*, transhumanists successfully lend an air of inevitability to the utopian scenarios they describe. The utopia of a radically enhanced post-humanity is presented not only as achievable, but also as the natural destination of a journey that humanity has been on right from the start. The future paradise is a paradise regained: the place where we are meant to be and where, precisely for that reason, we eventually *will* be. Paradoxically, becoming *more* than human is what *being* human is thought to be all about. "To choose to be better is to be human" (Savulescu et al. 2004, 670). We can certainly slow down the progression towards post-humanity, perhaps even suspend it for a while, but ultimately, we cannot prevent it. "We must progress on to transhumanity" (More 1996). The upshot is that we have no choice but to go along with the suggested transformation and, therefore, we can spare ourselves the trouble of thinking about whether we really *want* this to happen: it is going to happen anyway, so all resistance is futile.

Third, the strong emphasis on the wonders that allegedly await us in our future post-human existence makes our present condition appear far worse than it would if we did not compare it with an imaginary future in which all our dreams have come true. In comparison with such a future, our present life is bound to appear rather miserable. Think of Bostrom's claim, cited above, that each day will be a joy and that we will listen to music that is to Mozart what Mozart is to bad Muzak. The comparison effects a *conceptual devaluation* of the present. Mozart may not yet sound like bad Muzak to our merely human ears, but prompted by the mental image of the vastly superior music of the future, we have already begun to *think* of it as inferior. This conceptual devaluation of the present considerably increases the desirability of radical enhancement and a post-human future. In other words, the intended post-human condition does not appear so incredibly attractive because we find our present human condition so deficient, but rather we find the latter deficient precisely because, and to the extent that, the former is depicted in such bright colours. The brighter we make the future shine, the duller the present will appear.

The plausibility of transhumanist arguments concerning the desirability of radical human enhancement crucially hinges on utopian ideas that are deeply rooted in our cultural imagination and have a long history that links them to ancient dreams, hopes, and fears. We would do well to be aware of these roots, to reflect critically upon them and the ideals they promote, and to ask to what extent, and with what final purpose, transhumanist arguments tacitly appeal to particular conceptions of human nature and rely on deeply ingrained understandings of what we should strive to be and how we should act. It seems to me that if we look behind the arguments, it is not pure reason that we find, as transhumanists like to pretend, but rather a particular set of stories: about what it means to be human, what life is all about, and what the ultimate good is for beings such as us.

In the following chapters, I will not attempt to give an answer to these questions. Rather, I will reflect on what we are actually doing when we attempt to answer them.

Notes

1. For a fine account of Huxley's affinities to today's transhumanism, see Bashford (2013).
2. Nick Bostrom's personal website: www.nickbostrom.com. Retrieved in 2011, since then changed.
3. In his 2014 book *Superintelligence*, Bostrom warns against the dangers that superintelligent machines, whose imminent appearance he thinks is inevitable, will pose to the survival of humanity and suggests strategies to help us keep those dangers at bay. This does not, however, prevent him from concluding his investigation on an optimistic, crypto-utopian note, claiming that "our principal moral priority" today is "the reduction of existential risk and the attainment of a civilizational trajectory that leads to a compassionate and jubilant use of humanity's cosmic endowment." (Bostrom 2014, 260).
4. For a comprehensive critique of the transhumanist understanding of love and sex, please see my *Sex and the Post-human Condition*, Palgrave Macmillan 2014.
5. http://research.lifeboat.com/bostrom.htm
6. It is noteworthy that in his later book *Humanity's End* (2010), Agar abandons his earlier position and, without explaining his change of mind, now fully embraces the precautionary principle.

Bibliography

Agar, Nicholas (2004) *Liberal Eugenics*, London: Blackwell.

——— (2010) *Humanity's End*, Cambridge, MA: MIT Press.

Bacon, Francis (1924) *New Atlantis*, Oxford: Clarendon Press.

Bashford, Alison (2013) "Julian Huxley's Transhumanism", in *Crafting Humans. From Genesis to Eugenics and Beyond*, ed. Marius Turda, National Taiwan University Press, 153–167.

Berthoud, Gerald (2007) "The Techno-Utopia of 'Human Performance Enhancement", in: Utopie heute 1, ed. Beat Sitter-Liver, Fribourg: Academic Press/Stuttgart: Kohlhammer, 291–305.

Bostrom, Nick (2005) "The Fable of the Dragon Tyrant," *Journal of Medical Ethics* 31: 273–277.

——— (2009) "Why I Want to be a Post-human When I Grow Up", in *Medical Enhancement and Post-humanity*, eds. B. Gordijn and R. Chadwick, Dordrecht: Springer, 107–137.

——— (2010) "A Letter from Utopia", version 1.9: http://www.nickbostrom.com/utopia.pdf.
——— (2014) *Superintelligence*, Oxford: Oxford University Press.
Condorcet, Marquis de (1955) *Sketch for a Historical Picture of the Progress of the Human Mind,* London: Weidenfeld & Nicolson.
Coudert, Allison (1980) *Alchemy: the Philosopher' Stone*, London: Wildwood House.
De Grey, Aubrey, and Michael Rae (2007) *Ending Aging*, New York: Saint Martin's Press.
Eurich, Nell (1967) *Science in Utopia*, Cambridge, MA: Harvard University Press.
Fichte, Johann Gottlieb (1908) *The Vocation of Man*, London: John Chapman.
Gane, Nicholas/Haraway, Donna (2006) "When we have never been human, what is to be done? Interview with Donna Haraway", *Theory Culture Society* 23, 135–158.
Graham, Elaine L. (2002) *Representations of the Post/Human*, New Brunswick: Rutgers University Press,
Godwin, William (1971) *An Enquiry Concerning Political Justice*, London: Oxford University Press.
Halberstam, Judith, and Ira Livingston (eds.) (1995) *Post-human Bodies*, Bloomington and Indianapolis: Indiana University Press.
Haraway, Donna (1985) "A Manifesto for Cyborgs. Science, Technology, and Socialist Feminism in the 1980s", in: *Feminism/Post-modernism*, ed. L. Nicholson: New York and London: Routledge 1990, 190–233.
Harris, John (2007) *Enhancing Evolution. The Ethical Case for Making Better People*, Princeton: Princeton University Press.
Hauskeller, Michael (2007) "A Legacy of Love? Reflections on the New "Humanitarian" Eugenics," in *Utopie heute 1*, ed B. Sitter-Liver, Fribourg: Academic Press/Stuttgart: Kohlhammer, 405–419.
Hayles, N. Katharine (1999) *How We Became Post-human*, Chicago and London: The University of Chicago Press.
Hesiod (1993) *Works and Days*, Indianapolis: Hackett Publishing.
Huxley, Julian (1957) "Transhumanism", in Julian Huxley, *New Bottles for New Wine*, London: Chatto & Windus, 13–17.
Kurzweil, Ray (1999) *The Age of Spiritual Machines*, New York: Penguin.
——— (2005) *The Singularity is Near*, New York: Viking Adult.
Miah, Andy (2008) "A Critical History of Post-humanism", in *Medical Enhancement and Post-humanity*, eds. Bert Gordijn/Ruth Chadwick, New York: Springer, 71–94.

More, Max (1996) "Transhumanism. Towards a Futurist Philosophy", www.maxmore.com/transhum.htm.

Paul, Gregory S., and Earl Cox (1996) *Beyond Humanity*, Rockland, MA: Charles River.

Pearce, David (1995) *The Hedonistic Imperative*, www.hedweb.com.

Pico della Mirandola (1985) *On the Dignity of Man, On Being and the One, Heptaplus*, New York: Macmillan.

Pleij, Herman (2001) *Dreaming of Cockaigne*, New York: Columbia University Press.

Roco, Mihail C., and William Sims Bainbridge (eds.) (2003) *Converging Technologies for Improving Human Performance*, Dordrecht: Springer.

Savulescu, Julian, Bennett Foddy, and Megan Clayton (2004) "Why We Should Allow Performance Enhancing Drugs in Sport", *British Journal of Sports Medicine* 38: 666–670.

Searle, John (2010) *Making the Social World*, Oxford: Oxford University Press.

Stock, Gregory (2003) *Redesigning Humans*, London: Profile Books.

Trotsky, Leon (2005) *Literature and Revolution*, Chicago.

Val, Jaime de, and Stephan Sorgner (2010) "Metahumanist Manifesto", www.metahumanism.eu.

Wells, Herbert George (1967) *A Modern Utopia*, Lincoln: University of Nebraska Press.

Wills, David (2008) *Dorsality. Thinking Back through Technology and Politics*, Minneapolis: University of Minnesota Press.

Wolfe, Cary (2010) *What Is Post-humanism?*, Minneapolis: University of Minnesota Press.

3
Birds Don't Fly

So what *does* it mean to be human? Is there something like *the* human essence, something that we all have in common and that defines what we are? And will our radically enhanced descendants of the future still be humans—*better* humans no doubt, but still preserving the same human essence—or will they rather be emphatically *post*-human?

Any attempt to define the human essence has to deal with the undeniable fact of human variation. Everybody is different from everybody else. There are no two persons in this world that are completely alike. However, nor are they completely different. So for any two persons, there are dissimilarities and similarities between them. We acknowledge the fundamental dissimilarity by regarding and treating them as different individuals, and we acknowledge the fundamental similarity by regarding and treating them both as human beings. Of course, there may be other similarities; for instance, they might be both female or both male, both light-skinned or both dark-skinned, both blue-eyed or both brown-eyed, but in most contexts and for most purposes (not in all, though), the most *relevant* similarity between them consists in their both being human. Under normal circumstances, we have no difficulty recognizing each other as human and to tell humans and non-humans apart. This

M. Hauskeller, *Mythologies of Transhumanism*,
DOI 10.1007/978-3-319-39741-2_3

suggests that we must have some idea, however vague, of what makes a thing human. However, this idea may not always be the same, and not the same for everyone. In the late sixteenth century, Christian scholars were still debating whether women had souls and, consequently, whether they should be counted as human. Whole books were written on the subject, such as the anonymously published *A New Discourse Against Women, In Which It Is Proven That They Are Not Human* (Fudge 2006, 40). In the late nineteenth century, it was still common among white scholars to cast doubt on the humanity of black people and to denounce them as ape-like creatures or the missing evolutionary link between apes and humans (Wright 1998, 33). And perhaps we will soon enhance or change human beings in such a way that, after an initial phase of doubting whether the resulting beings are still human, we will not only accept them as human but also gradually come to see them as *paradigmatically* so. The other day, I had a discussion with my son Arthur (who is now eight) about birds and flying. Birds, he insisted, cannot fly, they only glide. When I pointed out that they by no means always glide, but also occasionally flap their wings to get up and stay in the air, he remained unperturbed. Yes, that was true, he said, but flapping one's wings is not flying, it is more like jumping in the air. Flying is what *planes* do. Planes fly, and *they* do not flap their wings. Hence, flapping one's wings is not flying. What I find remarkable about this argument is that, apparently for my son, the paradigmatic instance of "flying" is no longer what it used to be for everyone for thousands of years—namely, the flight of birds. Instead, it is constituted by the specific abilities and mode of operation of a human-made machine. Those machines *really* fly, whereas birds *merely* flap their wings. In the same way, the paradigm of what it means to be human may change over time when humans gradually change into, or get replaced by, their radically enhanced successors. What presently counts as human might then be regarded as merely subhuman or pre-human.

Now if that were to happen, would we then be justified in saying that this new race of people was *wrong* to regard themselves as human and us as subhuman or pre-human? Or would *they* be justified in saying that *we* were wrong to regard ourselves as human? This question is different from (though not entirely unconnected to) the one raised by Nicholas Agar (2013), whether the radically enhanced humans or post-humans of the future would be justified in assigning *a higher moral status* to themselves and whether they would thus qualify as post-*persons* in a moral sense. Agar has argued that it is not only possible but even likely that this would be the case. This would mean that they would be morally justified in sacrificing our lives and well-being not merely in cases of emergency, but also in cases of "supreme opportunities" (of which there will be many), that is, whenever such a sacrifice leads to "significant benefits for post-persons". For this reason, Agar believes, it would be morally wrong to allow any cognitive enhancement of people that might entail the risk of moral status enhancement. However, neither are there sufficient grounds to expect radically enhanced humans to have a higher moral status than unenhanced humans, nor would it, even if they did, be morally wrong to bring about their existence. We use moral status ascriptions mostly as convenient shorthand to indicate a difference in capacities that strikes us as morally relevant. Rocks have zero moral status because they cannot feel or think, so we cannot hurt or kill them. Whatever we do to them, it does not affect them. Animals, on the other hand, have some moral status because they can feel, so we can hurt them. What we do to them does affect them directly. However, we, or many of us, do not grant them full moral status because they lack certain other capacities that are connected to conceptual thinking, without which we believe (rightly or wrongly) an entity cannot really value its own existence. So we are allowed to kill them without needing a particular moral justification for it. Humans, on the other hand, feel and think, so both their well-being and their existence matters to them, and thus, we neither ought to hurt nor kill them. In all these cases, there is an intelligible connection between the abilities of an entity, the way and extent to which it can be affected by our actions, and the moral status that we grant it. The connecting element is the notion of harm, and it is hard to see how radically enhanced humans should be capable of being harmed in ways that we cannot. Agar seems to assume that we rightly claim to have a higher moral status than animals because we are so much smarter than they are, but what is morally relevant about our "practical rationality" or particular "cognitive power" (if anything) is not that we have more of it than animals, but instead that it is of a different kind. Following Locke's account of personhood, a person is "a thinking intelligent being that has reason and reflection, and can consider itself as itself", and according to Kant, it is an autonomous being, that is, one that is capable of acting out of respect for the moral law. Such beings are said to have an absolute value (instead of merely a price) and thus to be irreplaceable. In either case, you cannot have more or less of it. Either you can consider yourself as yourself, or you cannot, and either you are (in the Kantian sense) autonomous or you are not. So the kind of cognitive power that is morally relevant is, contrary to what Agar seems to assume, not something that can be improved upon. Hence, it is not conceivable that, if the improvement is big enough, this might result in a higher moral status. For all we know, there is indeed an upper limit to the capacities that determine moral status. Although it is correct that our idea of full moral status—the absolute value of human persons—seems to admit of degrees of violability, this is owed more to practical necessity rather than to any kind of theoretical commitment. We do not really think that some people are less violable than others—in fact, it is an essential part of our shared ethical self-understanding not to think that. There are, however, situations in which we have to make some tough decisions, where we cannot protect all and have to sacrifice some in order to save others. Making those decisions, though, does not imply any judgement about the moral status of those affected by them. It may well be that radically enhanced persons would appear less violable to us (and to themselves), simply because they are so amazing; that is, we might be more hesitant to sacrifice them in emergency situations. But from that alone, we cannot infer that they have a higher moral status. Even though we may prefer saving them rather than others, they are not entitled to it. It may sometimes be necessary to sacrifice an innocent person, but it is never right.

We may also doubt that moral status is an objective fact (as opposed to a mere ascription), which can then be discovered by beings with sufficient cognitive abilities. It may well be that beings that are vastly superior to us will be convinced that they are perfectly in their right to use us for their own benefit, but that does not mean that they know things that we do not. Consequently, there would be no reason for us to defer to their allegedly superior judgement (just as there would be no good reason for the cow to believe that it is perfectly all right for us to slaughter and eat her).

Agar concludes that cognitive enhancement should be kept within narrow bounds, because otherwise we are likely to face consequences that are "bad in moral terms". However, although we may have little reason to bring about the existence of a race of supermen (always assuming that they are not us, which is not necessarily the case), there does not seem to be anything *morally* wrong with it. If they, as Agar argues, are morally justified in using us for their own benefit, just as we are justified in using animals for ours, then there is no reason to think it would be morally wrong to bring their existence about—which, of course, in no way shows that we have a good reason to do so.

Now, although ascriptions of personhood often coincide with ascriptions of humanness, simply because for us the *paradigmatic* person is a human being, we can easily imagine non-human entities that meet the Lockean description of a person. Such entities could be, for instance, animals, machines, or alien life forms. Yet although it would then be difficult to contest that these entities are *persons* (in a descriptive, though not necessarily normative sense), we would probably still not want to call them *human* (although, for reasons that will become clear later, *they* might insist they are). Radically enhanced humans or post-humans, however, have good reason to see themselves (not as post-persons, but) as better humans and hence (if they understand their enhancement as an enhancement of the human *as* a human) as *truer* humans. If they made that claim and contested our claim to (full or real) humanness, could someone who is neither one of us nor one of them (say, an alien scientist from another planet) decide who is right and who is wrong? It seems to me that they could not. All that could be stated is that, on the basis of *our* own understanding of what it means to be human, we are clearly human, and on the basis of *their* understanding of what it means to be human, we seem to be not, or not entirely. In order to declare one of these contrasting understandings wrong, we would have to assume that the words we use always refer to the same basic property that makes a thing what it is, even when we have no idea what this property is, and consequently, never think of it when we use the term. In other words, we would have to assume that what we *think* a term means and how we, in fact, use it is one thing and what it *really* means and how it *ought* to be used quite another thing. Yet this assumption is hardly plausible. As individuals, we can of course be mistaken about the meaning of a certain term. I could, for instance, mistakenly believe that the word "pedestrian" refers to an adult male who has sexual relations with young boys, and use it accordingly. I would then be clearly wrong but only because this is not the way the word is *normally* used, that is, by the vast majority of speakers. Each of us can be wrong by violating the meanings established by a speech community, but the speech community itself can never be wrong because it defines, through the shared practice of speaking and writing, the very meaning of the words that individual speakers use.

The word "human" is no exception. As far as essence goes, the only essence we can ever hope to be relatively sure of is what John Locke called the *nominal essence* (1823, bk. 3, ch. 3, §15). Even if human beings had a "real essence" whereby they are what they are and whereupon their discoverable qualities depend, and even if we discovered this essence one day (neither of which is very likely), then we would have not much use for it because it would either be in accordance with the nominal essence of humanity, that is, with the abstract idea to which we attach the general name "human being", in which case we would not have learned anything new, or it would conflict with it, in which case we would not be willing to accept it. What we *mean* by the term "human" is not some hidden property in virtue of which we are what we are—Locke's real essence—but a loosely connected cluster of recognizable properties—Locke's nominal essence: "[T]o be a man, or of the species man, and have the essence of a man is the same thing. Now since nothing can be a man, or have a right to the name man, but what has conformity to the abstract idea the name man stands for; nor any thing be a man, or have a right to the species, but what has the essence of that species; it follows that the abstract idea for which the name stands, and the essence of the species, is one and the same" (§12).

Yet ideas change, and not least our ideas about what it means to be human, which means that the (nominal) essence of humanity also changes over time. Human nature then is constantly in a state of flux. From this, it follows that if two speech communities do not use the term "human" in the same way, then neither of them is right and neither is wrong.

However, we may still want to insist that whether or not people that are in some easily discernible way different from us are classified as human, or we are all classified as subhuman by future generations, is more than just a matter of linguistic convention. It should not matter how we classify, what we call human and what not, but to many people, it obviously does. Why is that so? Why do we care whether we are human or not, or someone else is? And why do we care what makes us human; that is, why do we care for the reason we call ourselves human? I think the answer to the first question (and thus, as we will see, also to the second) is that "human", to us, is usually more than just a descriptive predicate. It more often than not has a very strong prescriptive dimension. It is, just as the word "person" according to St Thomas Aquinas, a *nomen dignitatis*, that is, a title of honour, or a dignity-conferring name: "Persona est nomen dignitatis" (Thomas Aquinas 1980, lib. 1, dist. 10, q. 1, a5). Shakespeare's Juliet may have been right about roses when she declared that they would smell just as sweet when called by a different name, but that which we call a human by any other name would normally *not* smell as sweet. I am not saying that this is necessarily so. I am happy to concede that the term "human" can perhaps be used in a purely descriptive way, but my point is that this is not the way it is *normally* used. When it comes to applying the term "human" to an entity, we have obvious difficulties in separating the descriptive from the prescriptive. It is the same kind of difficulty that arises with respect to the term "art", which is also (often used as) a *nomen dignitatis*. The puzzled question "But is it art?" that many people cannot help asking when confronted with contemporary art clearly reflects their expectation that art should be something intrinsically valuable, worthy of being looked at, thought about, and being created in the first place. Thus, "Is it art?" usually means something like "Is it any good?" But it is not so much that we are *unable* to distinguish properly between the descriptive and the prescriptive, but rather that we are deeply *reluctant* to use the term in a purely descriptive manner. It would in fact be very easy to provide a definition of art if we could only bring ourselves to ignore the prescriptive aspect all together. "Art is everything that is regarded as art by at least one person who does not know the artist personally" is a perfectly good definition, but it is very unsatisfactory nonetheless because it does not give any indication why we should care for art. And that we *should* care is part of what we *mean* when we call something art.

With respect to the human, we have a similar reluctance to accept a purely descriptive definition. Questions such as "When does human life begin?" and "When does human life end?" are hotly debated. People clearly believe that it makes a huge difference in practical terms whether we say that human life begins with conception or that it begins with birth. Likewise, people believe that it makes a difference whether we say that human life ends with the irreversible termination of electrical activity in the neo-cortex, the irreversible termination of *all* brain activity, or the irreversible termination of all bodily metabolic processes (heartbeat and respiration). The reason for this is obvious: we tend to treat what we think of as humans in a way that is markedly different from the way we treat what we think of as non-humans (including not-yet-humans and no-longer-humans). And we reflect this experienced and expected treatment in our use of the word "human". When someone says that human life begins with conception, they usually mean to say that it is morally questionable to destroy a human embryo even at a very early stage of its development. Conversely, when someone insists that human life—or more precisely, the life of a human being—does not begin before birth, they usually take this to mean that abortion is justified or is at least not morally equivalent to killing a human being.

Equally hotly debated, and for similar reasons, are questions about the status of human–animal chimeras and hybrids. Is, for instance, a human–mouse chimera, that is, a mouse that contains some cells with human DNA, "partly human"? The answer depends on whether or not we think that our being human is a result of our genes, in the sense that this is "what makes us human", as James Watson once proclaimed (quoted in Nelkin and Lindee 1995).[1] But why should we care whether what technically speaking and from a biological standpoint is a human–mouse chimera is—or is not—classified as partly human? The answer is that being human, and perhaps even being "partly human", is associated with a particular moral status that is deemed considerably higher than the moral status of non-humans. Accordingly, a problem seems to arise as to how we should treat "partly human" mice: the same way as other mice, or like human beings, or something in between—better than other mice, although not quite like human beings? As the Scottish Council on Human Bioethics warned in a report issued in 2006, "[I]f an entity is accepted as having been created by human and non-human beings, then its whole identity and its entitlement to human rights and dignity could be challenged."[2] The reason for this is that the name "human" carries considerable moral weight, so we cannot call certain transgenic mice "partly human" without thereby suggesting that they have a higher moral status than other, "normal" mice.

This persuasive power of the term "human" is the reason why it does not only matter what is being called human and what not, but also how we justify applying, or denying, the term to an entity, or generally speaking, how we *define* being human. "Human" is one of those terms that, in the words of Charles L. Stevenson (1938, 333), have "both a vague conceptual meaning and a rich emotive meaning" and whose conceptual meaning "is subject to constant redefinition". It invites such redefinition precisely because of its rich emotional meaning, which is an ideal vehicle for advocating one's own ideas of what being human is about, how humans should be like, and how they should lead their lives, and for inserting these ideas into the very meaning of the word. "The words are prizes which each man seeks to bestow on the qualities of his own choice" (333). Stevenson points out that defining such a word is a way of redirecting interests. Each definition draws our attention to a particular aspect of the word or object that is meant to be defined and confers its emotional value upon this aspect. "Which meaning we choose, however, is no trivial matter; for we shall dignify that meaning by a laudatory title. To choose a meaning is to take sides in a social struggle" (344). I want to propose here that definitions of the human are as a rule, in Stevenson's sense, "persuasive definitions", that is, definitions that give "a new conceptual meaning to a familiar word without substantially changing its emotive meaning, and which is used with the conscious or unconscious purpose of changing, by this means, the direction of people's interests" (331).

Definitions can obviously have different purposes, but it seems that, independent of the purpose of defining "humanness", the ideal definition should refer to a property that is universal (so that all humans have it) and uniquely characteristic of humans (so that no non-human has it). Unfortunately, such a property is hard to find. As David Hull has pointed out, in "most cases, any character universally distributed among the organisms belonging to a particular species is also possessed by organisms belonging to other species; and conversely, any character that happens to be limited to the organisms belonging to a particular species is unlikely to be possessed by all of them" (Hull 1998, 383). Biological species are characterised by their very variability, without which they could not evolve. Yet, they do evolve, which makes it even more unlikely that we will find a characteristic that all humans and only humans at all times throughout the history of the species possess. It appears entirely impossible when we remember that, properly speaking, there is no such thing as the "history of the species" because there is no point in time where a particular species begins to exist and another point in time where it ceases to exist. The process of evolution knows no species boundaries. One species evolves into another without clear dividing lines. It follows, again, that if we are looking for a definition of the human, we must confine ourselves to the present and to what or who belongs to the human species, understood as a mating and reproduction network (Hull 1998, 384), *now*. If we do that and we try hard enough, we may come up with a property that only humans have.[3] The trouble is that this property might not strike us as particularly relevant. James Boswell, Samuel Johnson's biographer, once proposed (perhaps jokingly, but with Boswell you never know) to define man as a "cooking animal", reasoning that the "beasts have memory, judgment and all the faculties and passions of our mind, in a certain degree; but no beast is a cook" (Boswell 1924, 179, fn 1). This definition can of course be repudiated for neglecting the fact that not all humans can cook (a defect that is particularly common among males), but in *that* respect, Boswell's definition is no worse than many other definitions of the human that have been widely accepted and promoted for millennia, such as Aristotle's definition of the human as a *zoon logon echon*, or rational (reason- and language-having) animal. Clearly, not all of those who are biologically human are rational, or at least not more rational than some animals, and even the best of us are only partly rational, and often enough, decidedly irrational. Yet Aristotle's definition has been cited approvingly countless times, whereas Boswell's would convince no one but perhaps a chef.[4] Why is that so? The real problem with Boswell's definition is not that it is not universal enough but that it fails to highlight a property that can plausibly be understood as giving us that special worth that the attribute "human" seems to indicate.

The failure to lend support to this alleged worth is also the problem with Plato's famous definition of the human as a featherless two-legged being, which in itself may very well meet the requirements of a good definition. Perhaps all humans are indeed featherless and two-legged and perhaps no animal is. The rooster that Diogenes, the cynic, plucked and presented as "Plato's human" (Diogenes Laertius 1958, VI 40) presents no counter-example precisely because it first had to be plucked. But even if there were *in fact* other naturally featherless two-legged animals—animals that we would hesitate to call human—we *might* be the only ones. If we are, then Plato's definition of the human as a featherless two-legged being is as good as any other when it comes to finding a criterion to distinguish this particular kind of being from all other beings. Why then does it strike us as off the point, even as ridiculous? Because we usually demand more of a definition when we ask what a particular thing is. We want the definition to focus on what is essential about the defined. Yet how do we distinguish the essential from the non-essential? Why should featherless two-leggedness *not* be considered essential? I think that the main reason for this is that Plato's definition fails to capture the emotional significance conveyed by the term "human".

Every seriously proposed definition of the human entails and promotes some opinion about how humans should, and should not, be like. They are meant to persuade to a particular way of living. When we define ourselves with Aristotle as rational animals, or as political animals, we do not so much describe what we are, or what we think we are, but rather what we think we ought to be, or ought to aspire to be. The definition proposes an ideal of humanity that every individual human is called upon to validate. We can of course be irrational and we can, if we really want to, live in solitude and unconcerned about human relations, but the definition will tell us that this is not how we *ought* to live. We are told that being rational, or living and working for the community in the *polis*, is what being human is all about, and only if we view ourselves in this way and act accordingly, we are human in the full sense of the word. Thus, defining the human is always an attempt at defining the *truly* human. The definition is descriptive, but since what it describes is the truly human, which as a rule contrasts with the actually existing human, it is also prescriptive. Sometimes, it is even entirely prescriptive. Immanuel Kant thought that what makes us human is not rationality but *autonomy*, that is, the ability to act from duty, that is, purely out of respect for the moral law. However, he also declared that we cannot know whether there ever was, or ever will be, a human being that really acts autonomously. There may always be an ulterior motive, even when we sincerely believe that we act from duty alone. Yet even if humans might *in fact* be not autonomous, autonomy is still the hallmark of humanity. Kant's whole moral philosophy grounds in this paradigmatically persuasive definition.

The contrast between what humans actually are and what, by virtue of their humanity, they should be is also central to those definitions of the human that seem to draw a more realistic picture by focusing on the deplorable facts of human life. When Jonathan Swift redefines the human through the eyes of his Gulliver as a nasty, ugly, and brutish Yahoo, he forces his readers to look into a mirror in which they see themselves as they have never seen themselves before, showing them how far they are still away from realizing the ideal and becoming truly human.[5] Even Plato's seemingly silly definition of the human as a featherless two-legged animal can be seen in this light and used for expressing a notion of ideal humanness. Thomas Carlyle, for instance, repeatedly used it to express his contempt for a particular but very common type of human. In a letter to John Ferguson, dated 22 October 1820, Carlyle reports that he had been asked to work as a tutor for a young boy, but when he got there, he found that the boy was "a dotard, a semi-vegetable" and his "elder brother, head of the family, a two-legged animal without feathers, intellect, or virtue" with the "power of eating pudding, but no higher power" (Carlyle 1970, 285). And in his early novel *Sartor Resartus*, he mocks that, in the midst of human misery and violence, more than "five hundred thousand two-legged animals without feathers lie round us, in horizontal positions; their heads all in night caps, and full of the foolishest dreams" (Carlyle 1908, 16). This latter remark comes very close to a statement about human nature in general.

Yet all this pessimism about human nature does not in the least affect the *ideal*. On the contrary, it only serves as a reminder of the ideal and encourages us to renew our efforts to get close to it. Perhaps most people are in fact nothing more than two-legged animals without feathers that specialise in pudding-eating, but they should not be. There is (or should be) *more* to being human than that. Accordingly, when Prendick, the narrator of H.G. Wells' *The Island of Dr Moreau*, finally returns home and finds human company almost intolerable because he cannot help seeing the people around him as similar to the Beast People of Moreau's island, "animals half-wrought into the outward image of human souls" (Wells 1921, 173), he has lost his faith in humanity, but clearly has not given up his ideas about what it means to be (truly) human. He sees "faces keen and bright, others dull and dangerous, others unsteady, insincere; none that have the calm authority of a reasonable soul". The reasonable soul is what humans are supposed to have. It is the kind of soul that distinguishes them from animals. Because it is a human *proprium*, or rather *the* human *proprium*, it is proper for humans to be equipped with such a soul. Yet as it is, most humans bear only "the outward image of human souls" without actually having one. Prendick's reflections lead us back to the simile of the human soul that Socrates proposes at the end of Plato's *Republic*. There, we are asked to imagine the soul as a combination of three different entities: a giant multi-headed beast, a lion, and a human being, all bound together and furnished with the outward frame of a human being (588c-e). Only, explains Socrates, when the inner human being rules with the help of the lion over the multi-headed beast, then what appears to be human from the outside is also human from the inside. Then the Yahoo really becomes what he looks like to the casual observer: a human being. But in order for that to happen, a lot of work needs to be done.

The underlying notion of an ideal humanness, together with the pessimistic assessment that humanity, as it presently is, still has a long way to go to reach this ideal, points away from the present and towards a possible (and desirable) future actualisation of whatever is proposed as the defining aspect of our humanity. Hence, definitions of the human, whether they highlight a desirable and supposedly dignity-conferring property like rationality or autonomy or, to the contrary, a property that makes us ashamed of how we behave most of the time, often have a decidedly utopian character. Definitions of human nature are miniature utopias: they indicate a place that does not (yet) exist but that might and should exist in future. Thus, ironically, the hypothetical future race of radically enhanced humans, or post-humans, that I mentioned at the beginning of this chapter could base their claims to being the first true humans on our own definitions of humanness. They can claim to actually be what we only pretend to be or flatter ourselves to be. They can claim to be what Diogenes, the cynic—the same one that made fun of Plato—was looking for when he lit a candle in daylight and, on being asked what he was doing, replied that he was looking for a human being (Diogenes Laertius VI.14).

There are various, equally valid ways we can define ourselves. We can see ourselves primarily as rational beings, images of God, speaking, knowledge-seeking, art-producing, story-telling, tool-making, grave-digging, symbolizing, lying, laughing and crying, or self-designing beings, mortal or potentially immortal beings, natural or (partly) supernatural or "transanimal" beings (Jonas 1985). Not all of these properties are defining in the strict sense. Claims about human nature can either be about what is specifically human or what humans may share with other beings, for example, their mortality. We can say that mortality is part of human nature despite the fact that all living beings are mortal, in the sense that it is essential for the way we are and see ourselves as human beings, so an immortal human being would no longer be a human being at all. Alternatively, one could argue that in fact, only humans are mortal in the sense that only they (although again probably not all of them) are *aware* of their mortality. As Jorge Luis Borges once remarked: "To be immortal is commonplace; except for man, all creatures are immortal, for they are ignorant of death" (Borges 1962, 114). Yet whatever property we emphasise, whatever property, or complex of properties, we declare to be essential, our choice will make a difference also in practical terms. It will, as Stevenson said, redirect people's interests, or is at least intended to do so. And we *want* our interests to be directed when we seek a definition of humanness. We want to know what we are mainly in order to know what kind of life is the right one for beings such as us. We ask what it means to be human in order to find out what it means to be a *good* human. We seek to define ourselves, give ourselves an identity, and then try to be what we tell ourselves we are. The common and seemingly factual disagreement about what it means to be human is thus only the expression of a deeper disagreement about what the best way of being in the world as a human is. When we embrace a particular conception of human nature, we implicitly make a statement about what, in our view, human life is or should be all about, what matters or should matter in life, and what makes a human life good. In that sense, human nature is a myth, by which I do not mean that there is nothing that all humans and only humans have in common, but rather that each attempt at defining what we are is the telling of a story that implicitly or explicitly claims to be of prime significance for the way we ought to lead our lives. To a certain degree, all philosophies tell a story: about what it means to be human, about what is worth doing, desiring,[6] and fighting for, about good and evil and what life is all about. But the storytelling is most evident, or most concentrated, in claims about human nature.

In an article entitled "Science and Myth", John Maynard Smith points out that an important function of myths is to "give moral and evaluative guidance" and to provide "a source and justification for values", and that people repeat myths "because they hope to persuade others to behave in certain ways" (Smith 1998, 375 and 381). Definitions of the human are clearly myths in that sense. They are persuasive definitions that are meant to, and are taken to, provide moral guidance and justification for values, and they recommend that we behave in certain ways. Occasionally, this is openly acknowledged. Edward O. Wilson, for instance, the notorious champion of socio-biology or evolutionary psychology, admits in his book *On Human Nature* that scientific materialism is nothing but a mythology (Wilson 1978, 201), although he insists that it is a particularly powerful one because it "is the only mythology that can manufacture great goals from the sustained pursuit of pure knowledge" (207).

Incidentally, Wilson is one of those who believe that we can learn more about human behaviour, that is, why we do what we do, by looking at animal behaviour. This approach is thought to be justified by evolutionary theory. If the theory is right, then our behaviour must have evolved, which is taken to mean that it was naturally selected by virtue of its survival value.[7] This, however, might not be obvious anymore since today we live in a cultural setting that renders useless certain behaviour patterns that once might have been highly useful. Since our modern culture is, in evolutionary terms, very young, and our genetic composition, in which our behaviour is rooted, does not change that fact, we find ourselves behaving in a way that often seems to make no sense. Looking at analogous behaviour patterns in animals might then help us discover reason in the seemingly unreasonable by helping us understand why we do what we do. Understanding is important because only if we understand and acknowledge the powers that have shaped us can we hope to deal with them effectively. As Wilson says: "The learning rules of violent aggression are largely obsolete. We are no longer hunter-gatherers who settle disputes with spears, arrows, and stone axes. But to acknowledge the obsolescence of the rules is not to banish them. We can only work our way around them" (Wilson 1978, 119). This implies, of course, that we *can* work our way around them. We are not slaves to our genes in the sense that we cannot escape the behaviour patterns that we have inherited from our evolutionary ancestors. Often, however, evolutionary psychologists like Wilson are attacked for allegedly turning us into genetically determined natural automata, thereby denying human freedom of will and dignity and justifying political inequality, racism, rape, and several other morally obnoxious human dispositions. Even the title of Wilson's book, *On Human Nature*, used to provoke resentment because it underlines the socio-biological definition of the human as a *natural* being: an animal that differs in many ways from other animals, but an animal nonetheless. To some people, the expression "human nature" feels like an oxymoron, that is, a contradiction in terms, like "dry water" or "Christian science". That is because the word "nature" is regarded as signifying confinement, necessity, inevitability. In this sense, nature begins where human control ends. So if one believes that humans are basically free to do whatever they want to, all talk of "human nature" is inherently suspicious. However, denying human nature (in a narrow sense) is of course also a statement about human nature (in the wider sense). It is a different story that we tell about ourselves, a different way of making sense of what we are. To declare, as Pico della Mirandola did (1985, 3–7), that humans have no nature thus entails a positive claim about human nature. Put paradoxically, we could say that it is our nature to have no nature, and consequently, that we are nature-less animals or beings. This, of course, also has implications for human practice. In Pico's case, it was the claim that, since there are no natural boundaries to our being, we ought to take our destiny in our own hands and design ourselves. This claim is echoed by proponents of radical human enhancement today. We find it, for instance, in Gregory Stock's demand that we "seize control of our evolutionary future" and in the confident assertion that the "age of human self-design" has already begun (Stock 2003, 2).

Others, like the anthropologist Jonathan Marks (2009), seem to oppose the socio-biological naturalisation of the human because they fear the false moral conclusions that might be drawn from it. Marks argues that humans are essentially *cultural* beings, so we cannot learn anything from looking at animals or so-called primitive humans that live in an allegedly more natural setting. Humans, no matter where they live and how they live, are all equally cultural beings, equally far removed from nature. In his view, socio-biologists construct an "evolutionary origin narrative" that is designed to justify race-based and sex-based inequalities. Biology, he emphasises, "is irrelevant to the real issue of good and bad". Of course it is, and Wilson does not deny it. He does not commit the naturalistic fallacy that Marks accuses socio-biologists of. So what exactly is the problem? The problem is in fact the narrative that is built around (certain interpretations of) evolutionary theory—by proponents and opponents alike. If there is a problem at all, then it lies in what we make of the analogies between animal and human behaviour that socio-biologists are fond of pointing out. It all depends on whether or not we want to define ourselves as basically natural beings, and what we mean by it. So when Marks accuses socio-biologists of committing a naturalistic fallacy, he actually misses the point: biology may not be in itself relevant to what is good or bad, but when it is *thought* relevant, then it *is* relevant. It is a question of finding an image of ourselves with which we can identify. To understand ourselves primarily as essentially cultural, that is, non-natural beings, means telling, and listening to, a different narrative, but a narrative nonetheless. The socio-biological "nothing-butism" school, which according to Marks downplays or ignores what is uniquely human and "sees human behaviours as essentially unchanged ape behaviours", is a man of straw that does not really exist. But even if it does, then its ideological underpinning is easily matched by the ideology of what might be called "nothing-likeism", which downplays and ignores what we have in common with the great apes and other animals and declares that human behaviour has got nothing whatsoever to do with animal behaviour, not even the behaviour of our closest animal relatives.

The truth about our nature is very simple (and probably very unsatisfactory). People are all different from each other, and they are also very much alike. People also differ from animals in many respects, but again in many other respects, they are very similar to them. What is more important, the similarities or the dissimilarities? That depends on our interests. Nothing that we do is more natural than anything else. Nor less natural. Nothing we do is more human, or less human, than anything else. Everything we do is human, whether all humans do it or we are the only ones who do it, whether only humans do it, or animals, too. Whatever we do, we are always what we are, and we are *all* that we are. Hence, since all definitions are selective, no definition can capture what we are. In this sense, human nature is indefinable. Yet we may still need definitions of the human: not in order to learn what we are, but in order to make sense of it and to provide us with an ideal to live up to. Transhumanists have their own specific ideal of human nature. In the next chapter, we will have a closer look at it.

Notes

1. For a similar view, see Noonan (1993, 59).
2. A similar concern was raised by Robert and Baylis (2003): "All things considered, the engineering of creatures that are part human and part nonhuman animal is objectionable because the existence of such beings would introduce inexorable moral confusion in our existing relationships with nonhuman animals and in our future relationships with past-human hybrids and chimeras."
3. For the purpose of the argument, I am assuming here that being human and belonging to a particular biological species is co-extensive in the sense that if you are human, then you belong to that species, and if you belong to it, then you are human. However, we can imagine that the species evolves in such a way that it splits into two groups whose members can no longer reproduce with members of the other group. We would then have two biological species, but there is no reason why we should not regard the members of *both* of them as equally human.
4. I am, of course, exaggerating. There are sociologists who do take Boswell's aperçu quite seriously. See, for instance, Symons (2000).
5. For a more detailed account of Swift's understanding of human nature, see Hauskeller (2016).
6. Cf. Frye (1957, 136): "In terms of narrative, myth is the imitation of actions near or at the conceivable limits of desire. (…) The fact that myth operates at the top level of human desire does not mean that it necessarily presents its world as attained or attainable by human beings."
7. That this is at best only half of the truth and rests on a certain very questionable *interpretation* of the theory is pointed out by various critics, for instance, by Kitcher (1985) and Dupre (2001).

Bibliography

Agar, Nicholas (2013) "Why it is possible to enhance moral status and why doing so is wrong", *Journal of Medical Ethics* 39: 67–74.
Aquinas, Thomas (1980), ed. Charles H. Lohr, *Scriptum super Sententiis*, Fordham University Press.
Borges, Jorge Luis (1962) *Labyrinths*, New York: New Directions Publishing.
Boswell, James (1924) *Journal of a Tour to the Hebrides with Samuel Johnson, LL.D.*, ed. R.W. Chapman, London.
Carlyle, Thomas (1908) *Sartor Resartus*, London: Everyman's Library.
——— (1970) *The Collected Letters of Thomas and Jane Welsh Carlyle*, vol. 1, Durham, NC.
Diogenes Laertius (1958) *Lives of Eminent Philosophers*, London: William Heinemann/Cambridge, MA: Harvard University Press.
Dupre, John (2001) *Human Nature and the Limits of Science*, Oxford: Clarendon Press.
Fudge, Erica (2006) *Brutal Reasoning. Animals, Rationality, and Humanity in Early Modern England*, Ithaca and London: Cornell University Press.
Frye, Northrop (1957) *Anatomy of Criticism*, Princeton, NJ: Princeton University Press.
Hauskeller, Michael (2016) "Topsyturvy – Jonathan Swift on Human Nature, Reason, and Morality", in Janelle Poetzsch, *Jonathan Swift and Philosophy* (in press).
Hull, David (1998) "On Human Nature", in: *The Philosophy of Biology*, eds. David Hull and Michael Rose, Oxford: Oxford University Press, 383–397.
Jonas, Hans (1985) "Werkzeug, Bild und Grab", *Scheidewege* 15: 85–112.
Kitcher, Philip (1985) *Vaulting Ambition*, Cambridge, MA, London: MIT Press.
Locke, John (1823) *An Essay Concerning Human Understanding*, in The Works of John Locke. A New Edition, Corrected. In Ten Volumes, vol. II, London 1823.
Marks, Jonathan (2009) "The Nature of Humanness", in *Oxford Handbook of Archaelogy*, eds. Chris Gosden, Barry Cunliffe and Rosemary A. Joyce, Oxford: Oxford University Press, 237–53.
Nelkin, Dorothy, and Susan Lindee (1995) *The DNA Mystique: The Gene as a Cultural Icon*, New York: Freeman.
Noonan, John T. junior (1993) "An Almost Absolute Value in History", in: *Arguing About Abortion*, ed. Lewis M. Shwartz, Belmont, CA: Wadsworth, 51–59.

Pico della Mirandola (1985) *On the Dignity of Man, On Being and the One, Heptaplus*, New York: Macmillan.
Robert, Jason Scott, and Francoise Baylis (2003), "Crossing Species Boundaries", *The American Journal of Bioethics* 3/3: 1–13
Scottish Council on Human Bioethics (2006) 'Embryonic, Fetal and Post-natal Animal-Human Mixtures: An Ethical Discussion', *Human Reproduction and Genetic Ethics* 12/2: 35–60.
Smith, John Maynard (1998) "Science and Myth", in: *The Philosophy of Biology*, eds. David L. Hull and Michael Ruse, Oxford: Oxford University Press, 374–382.
Stevenson, Charles Leslie (1938) "Persuasive Definitions", *Mind* 47: 331–350.
Stock, Gregory (2003) *Redesigning Humans*, London: Profile Books.
Symons, Michael (2000) *A History of Cooks and Cooking*, Urbana/Chicago.
Wells, Herbert George (1921) *The Island of Dr. Moreau*, London: Heinemann.
Wilson, Edward O. (1978) *On Human Nature*, Cambridge, MA/London: Harvard University Press.
Wright, W.D. (1998) *Racism Matters*, Westport, Connecticut/London: Praeger.

4

Shitting Ducks

Karl Jaspers (1971, 26–27) once pointed out that all ideas of human perfection that we are so prone to construct are necessarily deficient: "We would like to see the human ideal. We would like to recognize in our thoughts what we ought to be, and what we can be on the basis of our obscure ground. It is as if in the represented image we were to find a certainty of our essence through the clarity of the idea of ideal humanity. But every conceptual and every visible form of being human lacks universal validity. The form is only one aspect of historic Existenz, not Existenz itself. And every form of possible human perfection proves upon reflection to be defective and unachievable in reality."

Nowadays, the most vocal proponents of the idea of ideal humanity are transhumanists. Transhumanism is less a philosophy than a *Weltanschauung*, a particular view of the world and our own place in it, according to which we, not as individuals but as a species, are destined to become far better than we currently are. This is more than just a possibility: it is what is *meant* to happen. And although it may not be entirely clear *what* exactly we are meant to be, it is pretty obvious that we are not there yet. Evolution has, as it were, still got plans for us. The general assumption is that what we *really* are is not what we are *now*. What we

M. Hauskeller, *Mythologies of Transhumanism*,
DOI 10.1007/978-3-319-39741-2_4

really are is what we can turn ourselves into. We are still growing up. The true human is still to be created. And it is to be created by us. We can, should, and will shape ourselves into what we have always meant to be, but never were. Modern, twenty-first-century technology will allow us to accomplish this goal and thus to fulfil our destiny as an ever-expanding, nature-defying, freedom-seeking race. And we don't really have a choice anyway, because, let's face it, the world as it is now is not really a place worth living in, at least not for beings such as us. As it stands, we have got too little control and we experience too much pain, our lives are far too short and generally rather miserable compared with what might be possible, and, worst of all, they will very soon end in death, the greatest of all evils. (And this is how we *know* that we are not where we are meant to be yet: this is so bad, it simply cannot be all there is to life.) In order to make living worth our while, we need to become radically different from what we are. The route to salvation is human bioenhancement, which is the improvement of human beings and ultimately human nature itself through biotechnological means.

This is where transhumanism channels mainstream culture. Human bioenhancement is a topic difficult to ignore these days. We stumble across it virtually everywhere we go and look. It is as if our whole world is about to turn transhumanist, if it hasn't already. There is hardly a journal or magazine that does not contain some report or at least some ad presenting us with a new technology that promises to make us better in some way or another. We are constantly asked to treat ourselves with, or support and welcome the development of, various anti-ageing devices, from anti-wrinkle creams that seduce us with names like *Forever Youth Liberator* (Yves Saint-Laurent) to yet-to-be-developed nanotechnological molecule repair units. We are encouraged to enrich or replace our bodies with various bits of machinery, to use mood enhancers and other feel-good drugs, intelligence enhancers, drugs that increase wakefulness and attention span, drugs that improve our memory and others that help us forget, and even morality pills that will help us not to abuse any of those wonderful new abilities that modern technology has allowed us, or will soon allow us, to acquire. Countless scientists are busy developing and refining the required technologies in order to justify all the hopes that the media have fuelled, and bioethicists do their professional best to convince us that all this makes perfectly good sense and is desirable and in fact absolutely necessary. Human nature is about to be changed. Of course, transhumanists and other proponents of radical human enhancement, and generally all those who still believe in progress with a capital P, will not be inclined to find this particularly worrying. On the contrary: the change in human nature that will or may occur as a consequence of the expected widespread use of certain enhancement technologies is not just a side product of the desired improvements. It is in fact its primary goal. This is because nature is often understood as a *limiting* force: it is what we call anything that sets limits to what we can do. According to this view, prevalent among transhumanists, it is not our *abilities* that determine our nature, but rather the *lack* of certain abilities: not what we *can* do, but what we can*not* do. We encounter our nature primarily in the form of boundaries, when we realise we can go no further and we simply cannot get what we want, not because the external world puts obstacles in our way, but because of ourselves, our own inability. Nature is not, as it was for Jaspers, the *encompassing*, the ground of our being, but something very tangible. Nature is the disease that prevents us from going to work and from enjoying life to the full. Nature is old age, which weakens us, and it is death, which puts an end to our life. Nature is the emotions we have, which we cannot fully control, and our relative lack of intelligence, which prevents us from understanding more than just a fraction of the world in which we live. Nature is our inclination towards evil, our moral defects. If that is how we look at it, then nature comes across as the chief enemy, namely as that which cannot be controlled. That is why any improvement of the human condition requires also a change of human nature, or more precisely, a restriction and curtailment of human nature, and ultimately its complete dissolution. Thus, nature must not only be changed. Rather, the hold that it has got over us must be weakened and, if possible, brought to an end. The enhanced human will not only have a nature that is different from ours. Ideally, they will have no nature at all, that is nothing that limits them in any way. The radically enhanced, post-human version of ourselves is envisaged as a natureless being. The nature of the enhanced human is in fact an unnature.

Yet our nature is very much identified with our body, that is, with the fact that our existence is, at least for the time being, inseparable from that of an organic body. For this reason, the attempt to overcome human nature is realised in practice as the attempt to reduce and ultimately eliminate our corporality. Thus, enhancing the human is often imagined as the merging of the human body with machines, the replacing of its organic and hence perishable parts with more durable and less easily destructible artificial devices, and finally the replacing or rather superseding of the organic body through the uploading of the individual person onto a computer, which would then allow us to lead a post-organic, digital existence. The nature of the enhanced human is, ideally, a bodiless nature, and for this very reason, an unnature, because it is the absence of a body which shows most clearly the absence of nature, or rather our liberation from it. Because our biological bodies are often perceived as deficient in various ways—they limit our freedom, are easily destructible, and condemn us to die—we look for an alternative way to exist, and we find it in the machine and its way of existing. Machines are attractive as a model for (post)human existence because they seem to allow an escape from the messiness of the human body. The more machine-like the human body becomes, the more it can be controlled and the more we make it our own by aligning the working of our bodies with our purposes. If the human body could be turned into (or be replaced by) a machine, we would finally be free to shape our own destiny. "I wish I were a machine", Andy Warhol once said, "I don't want to be hurt. I don't want human emotions. I've never been touched by a painting. I don't want to think. The world would be easier to live in if we were all machines" (quoted in Rorabaugh 2004, 201).

I have some sympathy for this desire. Who wouldn't? We all have bodies, or more precisely, we all *are* bodies—living bodies, feeling bodies, thinking bodies, but bodies nonetheless. And those bodies that we have and are, are obviously not perfect. There is a lot that they cannot do, which means that there is a lot that we, to the extent that we are those bodies, cannot do. We cannot, for instance, swim like fish, we cannot run like a cheetah, and we cannot fly. We may be able to think and understand things, but only as long as our brains are functioning properly, and there are a lot of things that we cannot understand, no matter how hard we try, and a lot of things that we forget, all because of the limited capacity of our brains. Here, too, our bodies are to blame. And even though we have built machines that more than make up for all these bodily deficiencies, that allow us to move in water a lot faster than fish, on land a lot faster than cheetahs, and even in the air a lot faster and higher than the most powerful bird, and others that allow us to think and remember things better and faster and perhaps more reliably, all these machines remain only surrogates that constantly remind us of what our bodies still cannot do. Notwithstanding all the power that these machines lend us, we may still envy the bird for its ability to fly, the fish for its ability to swim, and the androids of numerous science fiction novels and films for their ability to think. For we know very well that the power we have acquired through our machines is ultimately only a loan that we might be asked to return at any moment. The machines we use are, after all, external to our existence; they are not us, and we are very much dependent on their constant availability, which is largely beyond our control, at least as far as the individual is concerned. If the airport closes its doors, we cannot fly; if the car breaks down, we have to walk; if the internet connection goes down, we cannot think, and if there is an electric power outage or we lose our smartphones, *everything* breaks down and we instantly become completely helpless, so much are we used to rely on the functioning of the myriad of machines that surround us.

However, it has been pointed out to me (by Alexander Ornella) that precisely because the use of those machines is so essential to our existence that we cannot really do without them, we may just as well regard them as an extension of our bodies, that is, as an integral part of ourselves. In that sense, and from an anthropological perspective, machines *are* us. But be that as it may, the truth is that in our self-perception, we distinguish very clearly between our own body and the (other) machines that we use. I consider myself to be my hands, or in my hands, in a different (and much more intimate) way than I consider myself to be the computer or in the computer that I (or my hands) use to express my thoughts. And most importantly, we tend to compare ourselves, that is, our bodies and what they can and cannot do, with the machines that we use, and when we do so, the machines seem to come across as far more efficient and hence superior. Even if our body can be considered just one machine among others (or if those machines can be considered yet another part of our body), then that particular machine (or that particular part of our body) can still be distinguished from the others and be found, in comparison, deficient. And since we tend to feel a particular affinity to the machine that is our organic body, we recognise our apparent inferiority and feel ashamed. The German philosopher Guenther Anders first described this phenomenon more than 50 years ago in his seminal book *Die Antiquiertheit des Menschen* (The Obscolescence of Man), which, sadly, still awaits its translation into English.[1] Anders fittingly called this feeling of deficiency "Promethean shame" (Anders 1956).

Not only are our bodies, on their own, not very efficient, but they are also quite vulnerable and can easily be harmed or destroyed. Not only are there many things we cannot do, but there also many things that can be done to us. That is why we spend so much time and energy on finding ways to protect ourselves from a multitude of dangers: from other people, our own machines (cars, for instance), wild animals, natural disasters, and especially the many terrible diseases that our bodies are prone to developing. We—that is, each and every one of us—are in fact accidents waiting to happen. It is just a question of time before we run out of luck and succumb to one of those dangers. For no matter how hard we fight, we will eventually be brought down by our own bodies. There is, as it were, a traitor waiting within who will eventually open the gates to the enemy, to heart failure or cancer, to Alzheimer's or other forms of dementia, and this traitor is our body. Not very strong to begin with, it will gradually become weaker and weaker, less and less able to defend itself, until it fails us altogether and we die of some malfunction or other. Let's face it: human bodies are feeble, messy things. The ageing body, the diseased body, the body in decline only brings this essential messiness to the fore. It shows the world not only the actual state of a particular body and its eventual destiny, but also what our bodies have been all along, their very essence. Even the bodies of the young and beautiful, as immaculate as they may appear from the outside, are filled with things that we would rather keep concealed: with intestines, blood and other bodily fluids, waste products, and unpleasant smells. You really do not want to open the human body to see what is inside. Nor do you want to see most of what is coming out of it with depressing regularity. Although all these things may perform an important function and thus be part of a complex order, their overall appearance is that of chaos and dissolution. Their appearance serves to articulate the precariousness of bodily or rather fleshly existence, and thus foreshadows death.

Thus, it seems only too understandable that many people, to put it mildly, are not entirely happy with their bodies, or more precisely with the *kind* of body they have. It is not an individual problem, but a generic one. It is the current biological state of their existence that they resent. Undeniably, the idea of improving our bodies, to make them less messy, less vulnerable, or, preferably, of getting rid of them altogether and replacing them with something else—something better, more capable, and more durable—has a strong appeal, at least in the West.[2] Technologies that promise any of this are, by and large, enthusiastically welcomed by the consumer, and proponents of radical human enhancement are only too happy to point out how right we are to despise our messy, fleshly bodies. Only recently, Allen Buchanan attacked the occasionally expressed belief that the human body is a delicately balanced masterwork, in which everything fits together perfectly and works in complete harmony with each other. Nature, Buchanan reprimands us, is not at all a "master engineer", but more of a blundering blind fool whose products, far from being masterworks, are generally rather badly designed. Biological organisms, including the human, may be "finely balanced", but only in the way a house of cards can be said to be finely balanced. The slightest breeze can bring it down. This is, according to Buchanan, an intolerable situation. "If the human organism is so poorly designed as to be exceedingly fragile, then we may need to improve it if we are to survive" (Buchanan 2011, 158). Actually, we simply cannot afford not to do it: our very survival is likely to depend on it. Without enhancement, or more precisely without human bioenhancement,[3] human extinction looms.

Paradoxically, this argument, while on the one hand repudiating the idea that organisms have been optimally designed (by God or Nature) because, in truth, they have not been designed at all, on the other hand affirms the design category as a meaningful way to understand and evaluate the human body. By declaring the human organism to be "poorly designed", it is strongly suggested that we see it as an instrument, tool, or machine that performs a certain function and fulfils a certain purpose. We are encouraged to criticise the human body for the bad workmanship it exhibits, *as if* it had been designed with a particular goal in mind. Like a faulty machine, the human body fails to do what it is supposed to do, what it has been *designed* to do. In Buchanan's account, the human organism features as a survival-machine. And it fails with respect to the design goal of staying alive, not just for a while, but forever. Yet perhaps that is not our design goal at all. Perhaps we are not meant to last long, and not meant to perform certain feats that are presently beyond our reach. In that case, our bodies may work just fine. Or perhaps, more in accordance with contemporary evolutionary theory, there is no design goal whatsoever, and if there is not, then the human body cannot be properly understood as a machine at all. It may well be one in Descartes' or La Mettrie's sense as something that is entirely physical and that can be explained in its entirety without postulating a "soul" or some non-physical equivalent. However, it is not a machine in the sense of being inherently purposeful. The reason for the body's existence is not any particular function, which is always the case with machines (for instance, a coffee machine, which exists for the purpose of making coffee). Accordingly, lacking an inherent, in-built purpose, the human body cannot properly be said to have failed or to be deficient. However, even though our bodies in themselves may not have a purpose, that is, not one particular purpose, we as thinking, reflecting, and willing persons obviously do have purposes. For instance, normally, we desire to stay alive as long as possible, and we regret that our bodies do not seem to share this purpose. From this perspective, the problem is not so much that our bodies are imperfect machines; the problem is rather that they are not machines at all, while we wish they were. There is an obvious gap between what we want to do, or wish we could do, and what we actually can do due to the limitations of our bodies. In order to close this gap, we are supposed to turn our bodies into machines, that is, into something that completely serves our purposes. We need to control them so that they can no longer control us. As yet undesigned, they need to become designable. This means that machines are not what our bodies are, but what we want them to be. Today, *l'homme machine* is not a reality, but an ideal, something that we should aspire to become. Humans are fascinated by machines, love machines, envy machines, want to be machines, although they are also scared by the prospect. The machine is the better human, but also alien, not quite human any more. This ambiguity characterises most of the iconic stories about man-machines that are abundant in Western culture.[4] There is, to name but a few, the German writer E.T.A. Hoffman's beautiful automaton Olimpia, with which the hero Nathaniel, to his detriment, falls madly in love,[5] or the beautiful robot in Fritz Lang's *Metropolis*, meant to replace, or rather bring to life again in a more durable form, the dead wife of its inventor, the mad scientist Rotwang. More recently, there is the "Six Million Dollar Man" of 1960's television fame, who is rebuilt after a terrible accident, in the process of which several of his body parts are being replaced by "bionic" implants that vastly increase his strength and physical abilities, the cyborg policeman *Robocop* in Paul Verhoeven's 1987 film, the gradually humanised killing machine *Terminator* (aka Arnold Schwarzenegger in his most iconic role), the Cybermen in Britain's *Doctor Who*, and countless others. All these cultural images express ambivalence. It is the same ambivalence that feeds the debate about radical human enhancement, where attraction and fear can be found in equal measure, but even the attraction is only the flipside of a different kind of fear. The question is which fear will turn out to be the greatest: the fear of the machine, that is, of entering unfamiliar territory and perhaps becoming something that is no longer recognisably human, or the fear (and indeed revulsion) of the human, that is, of our own messy bodies and the pain and indignities and eventual annihilation that they have in store for us. From the fear of the human stems the attraction of the machine, which we haltingly, curbed by our fear of the unknown, approach.

It is not entirely clear, though, whether our fascination with machines has always been expressive of an underlying desire to *be* a machine, or at least more machine-like, or whether this is a more recent development. Ancient and medieval automata often imitated features of living beings, and the goal apparently was to make them as life-like as possible so as to trick the viewer into believing that the machine was in fact a real animal or human, or something very close to it, something that is alive and that acts of its own accord (Riskin 2003a). This tradition culminated in Jacques Vaucanson's famous automata, created in the late 1730s, the *Flute Player*, the *Tambourine Player*, and, the most impressive of them all, the *Digesting Duck*. What is interesting here is that the reason why the mechanical duck was seen as such a great triumph was that it was capable (or rather seemed to be capable) of a feat that can be almost paradigmatically associated with biological life, namely digestion. In plain words, not only could it eat and drink, but it could also shit. (In truth, of course, the food was not really digested, but instead replaced by already digested food that was hidden in a secret compartment.) A defecating machine was obviously as life-like as a machine could possibly get. Yet precisely for this reason, the mechanical duck seems to pose a problem for what I claimed earlier, namely that we are so much drawn to machines because they seem to present a welcome alternative to the messiness of our fleshly bodies. If that is true, then how can we explain that we celebrate a *messy machine* such as Vaucanson's shitting duck as a fantastic achievement? However, the whole point of Vaucanson's creations seems to have been to demonstrate that it was possible to make a machine that was, in all relevant respects, just like a living being. Because if it was possible, then this would show that the living body was, in fact, in all relevant respects like a machine. If it could be *imitated* by a machine, then it *was* a machine. Moreover, it was a machine that could, at least in principle, be controlled, designed, and redesigned. And if fundamental biological processes such as digestion are mechanically reconstructable, then they can also be controlled, and if they can be controlled, they are already a lot less messy.[6] Even the digesting duck presents an alternative to our biological existence. It makes it thinkable as a real possibility. But of course, we may also be genuinely interested in creating better machines, without intending to model ourselves on them: machines that are capable of doing things that only living beings, or only humans, can do. Looking back at the history of artificial intelligence from the late 1950s to today, it seems that the original motivation of those involved was to create machines that could actually think like humans, in the sense that they could take over certain tasks that required a certain amount of intelligence that as yet only humans seemed to have (Crevier 1993). The standard was human intelligence and the problem was how to get machines to achieve that standard. However, after the computer system *Deep Blue* managed to beat the reigning world chess champion Garry Kasparov in 1997, the attitude seems to have changed, so today we are more interested in figuring out how to allow humans to think, that is, to process and store data, as efficiently as a computer, rather than to find a way to make computers think like humans. The new standard is computer intelligence and the new problem is how to get humans to achieve that standard. Thus, the reference point for technological change is no longer the present constitution of humanity, but its (possible) future constitution. We have started to regard ourselves as unfinished business, as if the best was yet to come, as if we were still on our way to become what we are destined to be. And it is the machines, with their seemingly unlimited power, that show us the way and whose rapid progression prompts increasingly frantic attempts to catch up. The singularity is near, and we want to be ready when it hits us. Yet we are also afraid. Radical changes are not for the faint-hearted. In any case, we seem to be both repulsed and attracted by the machine. Yet although we both fear and love it, our love is greater than our fear because our fear of death is greater than our fear of change (for it is death that we ultimately seek to escape from by merging with the machine). Thus, we constantly make advances to the machine, though we proceed with a certain caution, or if not exactly caution, then doubt. We are not entirely sure of ourselves.

Now, I think we can distinguish four stages in which this cautious approach to the machine, this somewhat hesitant mechanisation of the human, takes place. I call those stages: (1) illusionism, (2) fortification, (3) replacement, and (4) displacement. The first two largely consist in time-honoured human practices, which, however, during the last century, have considerably gained in importance and scope; the third is relatively new, but already widely practised; and the last is, although seriously discussed by leading advocates of human enhancement, still science fiction and will most likely never be realisable, but as an ideal and a logical extension of what is already being done, it is very much alive.

Illusionism is the practice of changing one's appearance in order to accord with a commonly accepted standard of beauty or simply with what is deemed normal, or to render invisible, or less visible, the physical signs of ageing. I call this illusionism, not only because its purpose is clearly to create the illusion of youth, and that means, of being unaffected by the bodily decline that comes with getting older, but also because it is a kind of make-believe where we are asked to suspend our disbelief (or the knowledge that ultimately it is in vain), just as an illusionist persuades people to believe in magic. It is only rarely, or incidentally, the constitution of a particular body that motivates people to seek cosmetic surgery. They may not like their body as it is, but this dislike is just a reflection of the deeper dissatisfaction that comes with having a body, or such a "human, all too human" body, in the first place. We desire to change our bodies not so much because we are unhappy with how they have turned out (equipped with a nose that is too long or too short, or breasts that are too small or too big), but because we are unhappy with the body as such. Cosmetic surgery either aims at conformity, or at the opposite, at appearing to be something special. Those aims are not so far apart as it may seem. If my body conforms with the bodies of everyone else, if I no longer "stick out", then it no longer shows itself as a living body, which by its very nature is unruly and not adherent to a general rule or standard. Yet also when I try to make myself special by transforming my appearance in such a way that I cannot possibly be confused with someone else, I am signalling to the world and myself that the rules that govern everybody else's lives do not apply to me. I am different, my body is not messy,[7] I am making my own rules, I will not die. Likewise, when we try to hide the signs of ageing that our body exhibits, what we are trying to hide is the body itself, that is, its true nature. Living things rise and decline. They come into existence and go out of existence. That is why we wish for a form of existence that is not life, at least not life as we know it. It is to be a post-biological form of life, which is not really life at all, because life is defined by the fact that it assumes a precarious balance between being and non-being, that it is always on the brink of annihilation, that it constantly has to reassert itself, to make a stand against the "dying of the light" (Dylan Thomas, cf. Jonas 1966). To the extent that a post-biological existence would free us from death or the necessity of dying, it would also free us from the existence of a living being. We will be immortal machines. Cosmetic surgery is an attempt to reduce the messiness of biological life, to gain control, to halt ageing and decline. Sadly, however, it does not work, and we know it.

In her dystopian novel *The Year of the Flood* (2009), Margaret Atwood describes a place where women periodically go to get their ageing faces and bodies rejuvenated. They go there because they are frightened by the signs of mortality that their flesh exhibits. Yet when they come out, after all those signs have been removed for the moment, they are still frightened because they are already wondering when the whole thing might be happening to them again. "The whole signs-of-mortality thing. The whole *thing* thing. Nobody likes it, thought Toby – being a body, a thing. Nobody wants to be limited in that way. We'd rather have wings. Even the word *flesh* has a mushy sound to it." (Atwood 2009, 315).

We resent being just another thing in the world, with its implications of limitation, lack of autonomy and true agency, passivity, and ultimately, destructibility. We'd rather have wings, that is, some means to escape from our basic thingness, our being shackled to the material world, in which everything is subject to change and everything is bound to perish in the end. Of course, there is also a sense in which our bodies, made out of mushy flesh, are *not* things, a sense in which to be a mere thing actually appears preferable to being a living body. A thing may not be alive, but that also means that it cannot die, and if your fear to die is great enough, then you may prefer not being alive at all. (We will hear more about this in the last chapter of this book.)

But perhaps it is not so much death that we fear, but ageing, that is, the loss of our youth. "I hope I die before I get old", Roger Daltrey once sang with *The Who*, almost 50 years ago now, giving voice to the anti-establishment sentiments of a whole generation, but also to the fear that one day, one may end up being just like them, simply by growing up. I never really noticed there was a difference between the fear of death and the fear of ageing until I read what Atwood wrote next, following the passage quoted above: "If you really want to stay the same age you are now forever and ever, (…) try jumping off the roof: death's a sure-fire method for stopping time." (Atwood 2009, 316). Is this just a bad joke, or is there more to it, some keen insight into the nature of our desire to stay young forever? That desire is, after all, (by logical implication) a desire to stop time, and if you can only stop time (for yourself) by dying, then the desire to stay young forever is tantamount to a desire to die. It is a cleverly concealed death wish.

Unless, of course, eternal youth does not rule out change, so we could go on changing, gathering experience, co-creating ourselves and the world we live in, without having to age. I don't think that is possible, though, because it is not only our bodies that age. It is also our minds (Hauskeller 2011a). To not age, one would have to be like Peter Pan, the boy who refused to grow up, who every night forgets what he has experienced that day and who remains forever untouched by the events in his life, which means that he doesn't really have a life. And that is precisely why he ultimately represents death (as well as eternal renewal, which cannot occur without death). Atwood is right: eternal youth and death are one and the same, or more precisely: the one can only be gained if you are willing to pay the price of the other.

Fortification (physical enhancement): Another, perhaps more successful route to making bodies less messy, one that goes beyond appearance, is that of procedures intended to strengthen the body, to make it less vulnerable and more capable. This can be partly achieved through a healthy diet or exercise, or by training one's strength or skills. We get healthier, thereby slow down bodily and mental decline, gain more control over ourselves and our environment, become more independent, all of which increases our chances to live longer and better lives. Yet we are not entirely satisfied and secretly yearn for higher goals. We look up to the world's best athletes who have accomplished what we can only dream of. They seem to have mastered their bodies, turned them into machines fit for purpose, into running, swimming, and fighting machines. They are our heroes, our gods. They appear invincible, or at least, once again, let invincibility appear a real possibility. Yet they also show us, time and again, the limits of the human body. We can work on our bodies all we want, but as long as their basic constitution remains unchanged, they will still fail us in the end. Even the gods of sport get older and lose their power. There always comes a day when even the seemingly invincible are being defeated by others, who replace them as the objects of our admiration. So we need to go beyond the biological body and pursue a different kind of fortification that points beyond what is humanly possible and brings us closer to an actual merger of human and machine. Just like a medieval knight could use an armour to protect him from being impaled by his enemies in battle all too easily, we can now use artificial exoskeletons, such as those developed by the Japanese company *Cyberdine*, which considerably augment the body's physical strength and durability and are principally designed to make soldiers more efficient, but can also be used for more peaceful purposes like cleaning up the mess after the meltdowns in Fukushima. The movement of the exoskeleton is controlled by a computer that detects and translates muscle movements via a sensor on the bearer's skin (Greenemeier 2011) and is thus more than just an external device that is consciously operated by its bearer. It is more like an extension of our body, but not in the way that a sword or a shield might be seen as an extension of one's body. We do not actually have to use our full muscle power to operate it, but almost, though not quite, just think about moving and the machine does the rest, thus effectively integrating a power into our body that is not our own. Brain–computer interfaces (BCIs) have a similar function, though they go a step further by connecting a device-operating computer directly to the brain or other parts of the central nervous system. BCIs could help people with locked-in syndrome to communicate, but can also be used in computer gaming or in the military for controlling weapons and other machines by thought alone. Here, the distinction between human and machine gets even more blurred. Just as I control my biological body (or parts thereof) through my will, BCIs allow me to control a machine in exactly the same way. The machine thus functions as an artificial body; that is, it becomes, for all intents and purposes, *my* body. While I retain my biological body, I add another, artificial, more powerful and capable body for me to use, and to identify with. Thus, it allows the kind of radical transformation that we are familiar with from numerous twenty-first-century blockbusters featuring superheroes, who, as Heimerl (2015) has pointed out, overcome the deficiencies and vulnerabilities of their own human bodies by donning a costume or armour that covers and conceals them and effectively transforms them into an entirely different kind of being, namely a superhero commanding a superhero's sweat- and blood-free uber-body.

Replacement: BCIs supplement the biological body with an artificial one that can do things that the biological body cannot do. There may still be things that the biological body can do that the artificial body is not able to do, at least not yet, but there does not seem to be an *in principle* reason why this should *have* to be the case. Once we have figured out how to build devices that perform the same function as the various parts of the biological body, there will be no reason, it seems, to retain the biological body. We can now start replacing those parts whose function can already be assumed by a machine and then gradually replace more parts as we go along. This seems to make sense for those parts that have ceased either to function completely or to function as well as we find desirable, or that threaten to stop functioning properly sometime soon, which is in fact true for almost *all* parts of the biological body. This is actually what we do with machines: it is called maintenance, and we do it to keep them running properly. The process of the gradual cyborgisation of the human has, of course, already started. More and more parts become replaceable, and the pressure and willingness to actually replace them increase steadily. We have, in our own view of ourselves, become 'men of parts,' fragmented, the Lego version of a human. We have become used to neuroprosthetic devices such as cochlear or retinal implants that replace our sense organs. Artificial lungs, hearts, and even brains may soon be available for everyone and replace our vital organs. We could also custom-tailor our bodies for all sorts of purposes, including the most trivial ones, like the New Jersey artist who made headlines when he had four magnets implanted in his arm to hold his iPod. Artificial limbs are no longer a poor substitute of the real thing, as they used to be even a decade ago. Instead, they appear increasingly desirable also to those with perfectly healthy limbs. They have become an object of envy, a must-have for the technophile in us. Hugh Herr, director of the Biomechatronics group at the Massachusetts Institute of Technology (MIT) Media Lab, designs bionic limbs for amputees that are meant to feel so natural to their owner that they become a genuine part of their identity, while being far better than the original, so there is nothing to be missed. An amputee himself, he declares that he would not want his old legs back. Losing them provided him with the golden opportunity to create a better body for himself and others. His goal is to gradually "rebuild the human from the ground up", literally, starting with the ankles, then the knees, the hips, and all the way up to the top, which has the great advantage of making the body indefinitely "upgradeable": "So every few months, I get a hardware and software upgrade. And as my biological body ages, my artificial limbs get better and better" (Gupta 2012). While the biological human declines, the machine-human moves up and forward.

Displacement: However, even the most durable and capable body is still a body. It may not age, but it can sure be destroyed by natural forces. Although much less messy, it still retains some essential messiness, like a mechanical duck with a digestive system. Thus, from the point of survival, the goal has to be to get rid of the body altogether, and once and for all. This can only be achieved by the as yet still future technology of mind uploading, which would allow us to give up materiality all together and to acquire "digital immortality" (Sandberg and Bostrom 2008, 5). Whether that will ever be possible is highly doubtful, mostly for philosophical reasons (Hauskeller 2012). But the point I am trying to make here is simply that the ultimate solution to what we could call the "messy body" problem is the dissolution of materiality. To get rid of messiness, we need to become machines, but in the last resort, as Ray Kurzweil put it, "spiritual machines".

Do we have to, though? And should we? I don't know, and I have here deliberately avoided making any ethical assessment of the whole development. However, it seems clear to me that we can look at our biological body in more than one way. We can hate and despise it, obviously, but we *can* also love and admire it. One way of looking at it reveals its many defects. From that perspective, the body is something that is *essentially* messy and that, for this reason, needs to be overcome. This view betrays an almost Manichean understanding of our bodily existence. The body is bad; salvation (and true human identity) lies elsewhere. However, there is a different perspective on the body that is equally justified. Even though our body prevents us from doing certain things, it also allows us to do many other things as well. The truth is that is not possible to be capable of everything. Any ability is also a limitation. What enables us to do one thing might prevent us from doing other things. *Omnis determinatio est negatio*, as Spinoza put it: all determination is a limitation. It involves an exclusion of things that one is not. From this point of view, the desire for omnipotence, to have no limits, is an unwholesome fantasy. Why not rejoice in the things that we *can* do instead of deploring those we cannot do? Our biological, human, messy bodies are in fact facilitators. They are essentially enabling by allowing us to experience beauty, love, pleasure, colours and sounds, and—yes—thoughts, too. This of course does not mean that we should not try to enhance our human bodies in any way. However, if we adopt a more positive attitude to our body, we might find that the whole enhancement enterprise, which I have here interpreted as an attempt to turn our bodies into machines (spiritual machines eventually), loses a lot of its urgency. There appears far less to be gained. Of course, ageing and death can cause considerable misery, but they also allow life to continue in new ways and fresh eyes to see the world (Hauskeller 2011). All things considered, they are fine, an acceptable price to pay for the opportunities that life holds in store for us, not least because of the kind of bodies we have, and it is certainly no solution to get rid of the body altogether. The glass may be half empty, but it is also half full, or rather full enough. Or simply full: full of opportunities. A healthy body is a great gift—something to treasure, certainly to protect, but not something to get rid of and replace as quickly as possible.

There is, however, another understanding of human nature that is just as important for transhumanism and generally the project of human bio-enhancement as the one we have just discussed. This other nature will be the subject of the next chapter.

Notes

1. Although an annotated translation of the first, central section of the book, the one on Promethean shame, has just been published by Christopher John Müller (2016).
2. Paul Gilbert (2015) has suggested that the invulnerability of the body is, in fact, a distinctly Western ideal. Western drone attacks exemplify this ideal insofar as the attacker remains safely out of harm's reach, whereas the (Eastern) suicide bomber accepts the destruction of his own body as something that ultimately cannot be avoided.
3. It goes without saying that, in a certain sense, we have always enhanced ourselves—namely, with tools and machines—without the use of which human survival would not have been possible. What Buchanan claims is that this kind of enhancement is no longer sufficient, that we need to extend it to our own natural constitution to have any chance of survival.
4. For an historical overview, see Wood (2002) and, emphasising the erotic aspects of the man–machine relationship, Ferguson (2010).
5. For a more detailed discussion of Hoffmann's tale, see Hauskeller (2014, ch. 3).
6. Cf. Riskin (2003b, 622): "Defecation and chess playing had something in common: both seemed beyond the bounds of mechanism and thereby provoked mechanicians who were interested in testing the limits of their craft to become conjurors."
7. I suppose it is theoretically possible to deliberately increase one's body's messiness, for instance, in order to oppose cultural norms and oppressive ideals that demand the opposite from us. Thus, the feminist writer and activist Kathryn Morgan (1998, 278–9) once suggested that women should use cosmetic enhancement technologies to make themselves *uglier* (by, for instance, using wrinkle-*inducing* creams and having their breasts surgically pulled down).

Bibliography

Anders, Günther (1956) *Die Antiquiertheit des Menschen*, Munich: C.H. Beck.

Atwood, Margaret (2009) *The Year of the Flood*, London: Bloomsbury.

Buchanan, Allen (2011) *Beyond Humanity?* Oxford: Oxford University Press.

Crevier, Daniel (1993) *AI: The Tumultuous Search for Artificial Intelligence*. New York: Basic Books.

Ferguson, Anthony (2010) *The Sex Doll. A History*, Jefferson & London: McFarland & Co.

Gilbert, Paul (2015) "'You love life, but we love death': Suicide Bombings and Drone Attacks" in *Making Humans*, ed Alexander Darius Ornella, Oxford: Inter-Disciplinary Press.

Greenemeier, Larry (2011) "Robotic Exoskeletons from Cyberdyne Could Help Workers Clean Up Fukushima Nuclear Mess." *Scientific American* 9/11/11 http://blogs.scientificamerican.com/observations/2011/11/09/robotic-exoskeletons-from-cyberdyne-could-help-workers-clean-up-fukushima-nuclear-mess/.

Gupta, Sanjay (2012) Interview with Hugo Herr. *CNN Transcripts* 25/3/12 http://transcripts.cnn.com/TRANSCRIPTS/1203/25/nl.01.html.

Hauskeller, Michael (2011) "Is Ageing Bad for Us?" *Ethics and Medicine* 27/1: 25–32.

——— (2011a) "Forever Young? Life Extension and the Ageing Mind", *Ethical Perspectives* 18/3: 385–406.

——— (2012) "My Brain, my Mind, and I. Some Philosophical Assumptions of Mind-Uploading." *International Journal of Machine Consciousness* 4/1: 187–200.

——— (2014) *Sex and the Posthuman Condition*, Basingstoke: Palgrave Macmillan.

Heimerl, Theresa (2015) "Ascetic Athletes: The Hero-Body in Contemporary Hollywood Cinema", in Anthropotechnologies, ed. Alexander Darius Ornella, Interdisciplinary Press.

Jaspers, Karl (1971) *Philosophy of Existence*, Philadelphia.

Jonas, Hans (1966) *The Phenomenon of Life*, New York.

Morgan, Kathryn P. (1998) "Women and the Knife: Cosmetic Surgery and the Colonization of Women's Bodies", in: *Sex/Machine: Readings in Culture, Gender, and Technology*, ed. P.D. Hopkins, Bloomington: Indiana University Press, 261–285.

Müller, Christopher John (2016) Prometheism. Technology, Digital Culture, and Human Obsolescence, London/New York: Rowman & Littlefield.

Riskin, Jessica (2003a) "Eighteenth-Century Wetware", *Representations* 83: 97–125.

——— (2003b): "The Defecating Duck, or, the Ambiguous Origins of Artificial Life", *Critical Inquiry* 29/4: 599–633.

Rorabaugh W.J. (2004) *Kennedy and the Promise of the Sixties*, Cambridge: Cambridge University Press.

Sandberg, Anders, and Nick Bostrom (2008) *Whole Brain Emulation. A Road Map*. Oxford: Future of Humanity Institute, Oxford University.

Wood, Gaby (2002) *Living Dolls. A Magical History of the Quest for Mechanical Life,* London: Faber & Faber.

5

Stealing Fire from the Gods (and the Weak)

As we have seen in the previous chapter, the term "nature" is generally used by transhumanists to refer to anything that limits us in any way (and that, for this reason, needs to be overcome). This understanding of nature has a decidedly negative connotation. Yet there is also a *positively* connoted concept of (human) nature, according to which nature is not that which needs to be overcome, but rather both that which *enables* us to go beyond those "natural" limitations and that *for the sake of which* we should go beyond them. We turn against nature as limitation, as we must precisely because it is our nature to do so. As Gregory Stock (2003, 2) once put it, stealing fire from the gods "is too characteristically human". And because we can only protect, or perhaps honour, this nature if we do everything in our power to resist that other, limiting, nature, we also have every right to do so. According to this understanding, man is, as Nietzsche said (1966, WII, 623), the "as yet undetermined animal" ("das noch nicht festgestellte Tier"), but not so much in the sense that we would require another thing to complete ourselves and become fully determined (be it society or technology or something else), but rather both in the sense that we possess *possibilities of being* that no other animal has and

M. Hauskeller, *Mythologies of Transhumanism*,
DOI 10.1007/978-3-319-39741-2_5

that perhaps have never been realised by any human yet, and in the sense that we fulfil our human destiny in the pursuit of those possibilities.

Our being-as-yet-undetermined is not so much a fact of human existence that we have to cope with in one way or another, as an essential possibility, but then again, not merely a possibility either, but also a *mission*. Pico della Mirandola (1985, 4) already described the human as an animal whose nature it is to have no nature, and believed that this was exactly what made the human special, what gave us dignity. Reaching from the lowest to the highest, all spheres of being are open to us, but there is no doubt that in order to fulfil our destiny and become truly human, we need to aspire to the highest. Potentially, we are all gods, and because we are and to the extent that we are, we *ought* to be gods, so we fail to be what we are (or meant to be) if we content ourselves with being animals.

A very similar and equally normative understanding of human nature often underlies current demands for a biotechnological enhancement of the human. Nothing seems to prevent us anymore from designing ourselves any way we see fit. That we forego the possibility of bettering ourselves (i.e. of overcoming the current limitations of our existence) is hardly imaginable, not only because we are constituted in such a way that we find ourselves pushing ever forward, but also because we would betray our own nature if we did. We can think. We can judge. We can take control of things in accordance with our thoughts and judgement; we can shape the world, and shape and reshape ourselves. This kind of creative engagement with the world, the reshaping of the given, is the true goal of the rational faculties that we possess, and it is this goal that makes us what we are. So, in this view, it is not the purpose of reason to enable us to admiringly contemplate "the starry heavens above me and the moral law within me" (Kant). Rather, we have reason so that we can use it to *improve our lives*—and ultimately the best way of doing that is by improving ourselves (Hauskeller 2009). Thus, human reason is primarily not a tool for the construction of theories about the world, but essentially and eminently *practice oriented*, and it is our ability to live by this reason and to give it as much room as possible that marks us out as humans and makes us special.

Pico, however, thought that the kind of improvement that reason was to serve was primarily a *moral* improvement, a realisation of man's higher nature. Today, this is no longer the case. On the contrary: if we take "human enhancement" to mean any *particular* improvement that might *result* from the general dislimitation and the unlocking of new possibilities, then there is no human enhancement, simply because there is no particular improvement that is being sought. The main object seems to be freedom *itself*, and not necessarily the freedom to reach certain goals that we have identified as desirable but as yet have not been able to reach because we have been prevented to do so by the limitations of our nature. The real object of desire seems to be not the possibility to do or be *this* or *that*, but rather to do or be *anything* that we might wish to do or be, whatever it is. Thus, freedom from determination is not primarily a necessary means to reach certain ends. It *is* the end. Even when other, more concrete goals are pursued, they are ultimately seen as means to achieve greater freedom.

By way of an example, let us have a look at radical life extension and the defeat of death (or more precisely, the necessity of dying), which appear particularly urgent to some of the most vocal proponents of radical human enhancement.[1] Countless scientists are busy trying to figure out what exactly makes us age, in the hope that they might find ways to slow down and halt ageing, and that one day, we may even be able to reverse it. This, however, can only be achieved if we manage to re-programme the human body, without which we cannot yet exist and whose constitution prevents us from attaining those goals. We need to change our bodily processes in such a way that a free space ensues, a, as it were, natureless space, which allows us to live on indefinitely. But if you ask *why* we should want to live so long, what a radically extended life span is *good* for, then more often than not, you will be told that we need a longer life in order to be able to realise the many possibilities of our existence.[2] Hence what immortality promises is relief from the necessity to commit oneself to a particular way of life, or more precisely, to being a particular person who, with increasing age, finds it more and more difficult to depart from their well-trodden life path and to radically reinvent themselves. It is commonly thought that a potentially unending life would allow us to start over whenever we wish to do so. We could shed our old lives like a snake sheds its skin, to emerge rejuvenated both in body and in mind. That this is in fact an illusion, I have argued elsewhere (Hauskeller 2011a). The point I am trying to make here is simply that it is ultimately indeterminacy itself, the being-undetermined, that is regarded as intrinsically valuable and as being the normative core of human nature. We are "undetermined animals" not in actual fact, or at least not sufficiently so, but with respect to our inherent potential and ultimate purpose. In other words, it is not the human as he or she is today who is undetermined, but it is the radically *enhanced* human who is, or will be, and it is precisely the expected decrease in determinacy that makes him an enhanced human, that is, a *better* human. And if we understand determinacy as nature, or nature as determinacy, then the nature of the radically enhanced human really is, as noted above, his unnature.

This is very different from what Nietzsche had in mind when he defined human nature as "not yet determined". There is in fact little that mainstream transhumanism shares with Nietzsche's philosophy other than a few key terms. By mainstream transhumanism, I mean the kind of transhumanism that is publicly promoted by the more prominent academic figureheads of the movement. As we shall see at the end of this chapter, not all transhumanists stick to the official party line. By and large, however, Nick Bostrom (2005a, 4) was spot on when he maintained that there are only surface-level similarities between Nietzsche's vision of the Ubermensch (overman, or overhuman) and the transhumanist conception of the post-human, even though others have later argued that the similarities are in fact "significant" and can be found "on a fundamental level" (Sorgner 2009, 29). Max More even admitted to have been strongly influenced by Nietzsche, which then led to the publication of his seminal essay "Transhumanism: Towards a Futurist Philosophy" in 1990 and inspired the very *name* of the movement: transhumanism (More 2010, 2). Without Nietzsche, then, there would arguably not have been a transhumanist movement. Yet, although this might be historically correct, there are in fact some essential differences between the standard transhumanist and the Nietzschean mythology that should not be overlooked.

First of all, transhumanists generally believe that it is both possible and desirable to improve human nature by means of technology (More 2009). They tend to assume that by "making better people" we will, as John Harris (2007, 3) puts it, make "the world a better place". Posthumans will allegedly lead happier, more fulfilling lives than we do now. This assumption is the main reason why transhumanists demand that we pave the way for post-humanity. In other words, there is a moral imperative at the heart of the transhumanist agenda. David Pearce calls it the "hedonistic imperative"—lifelong well-being as a basic human (and nonhuman) right—while Julian Savulescu (2001) calls it the "principle of procreative beneficence" which, if adhered to, naturally leads to the embracement of radical human enhancement, and per implication, post-humanity.[3] Nietzsche, on the other hand, had nothing but contempt for those who sought to improve the human condition, such as John Stuart Mill, whom he denounced as a "blockhead" (Flachkopf) because Mill still believed in good and evil (both natural and moral) and felt that one should make it one's duty to bring about the victory of the former and the destruction of the latter (1966, WIII, 665). According to Nietzsche, the philosopher needs to position himself "beyond good and evil" because there are no moral facts and nothing that is truly better or worse than anything else. Happiness, for instance, is not to be considered better than suffering. To believe otherwise indicates a grave error of judgement. And more than that: trying to improve humanity is actually an attempt to "suck the blood out of life", an act of "vampirism" (1966, WII, 1158). Consequently, Nietzsche fervently denied that he himself intended any such thing: "The last I would promise is to better humanity." (1966, WII, 1065).

Transhumanists may want to revaluate certain aspects of our existence, but most transhumanists (with, as we will see later, one notable exception) do not, as Nietzsche did, advocate the revaluation of all present values. On the contrary, they emphasise the continuity between (past and present) humanist, (present) transhumanist, and (future) post-human values and see themselves as defenders of the Enlightenment's legacy against its modern (bioconservative) enemies. "The posthuman values", writes Bostrom (2005b, 5), "can be our current values". Of course, a few things that are supposed to be valuable by some, such as the "natural", are discarded, but on the whole, a transhumanist would regard as good and valuable what is commonly regarded as good and valuable, for example, a long, healthy, and happy life, intellectual curiosity and proficiency, the ability to form deep and lasting relationships, and so on. Nietzsche, on the other hand, wanted to turn our whole system of values upside down, or rather, rip it apart. He prided himself to be the "first immoralist" and hence "destroyer par excellence" (1966, WII, 1153). What was commonly regarded as evil needed to be recognised as the highest good. "Evil is man's best power (…) necessary for the best of the overhuman" (1966, WII, 524). He wondered whether not all great humans were in fact evil (1966, WIII, 449), and he specifically and repeatedly mentions Cesar Borgia as "a kind of overhuman" (1966, W2, 1012), whom he admiringly describes as a "human predator" (Raubmensch) (1966, WII, 653). Compassion, charity, loving one's neighbour—traditional Christian values, but not alien to transhumanists either—are scoffed at as symptoms of decadence. According to Nietzsche, universal altruism would take the greatness from existence and effectively castrate humanity (1966, WII, 1155). Consequently, what puts Nietzsche's (or more precisely, Zarathustra's) overhuman over the merely human is precisely his indifference to common moral concerns: "the good and just would call his overhuman devil" (1966, WII, 1156). Most transhumanists would not want to hold that the post-human is post in this respect.

Transhumanists continue the logocentric tradition of Western philosophy. By and large, they believe that what makes us human, and what is most valuable about our humanity, is the particularity of our minds. We are thinking beings, conscious of ourselves and the world, rational agents that use our environment, including our own bodies, to pursue our own freely chosen ends. And because our essence consists in our thinking, it is at least conceivable that we may one day be able to transfer ("upload") our very being to a computer (or another biological brain) and thus achieve some kind of personal immortality. Generally, the organic body is held to be replaceable. Nietzsche, however, opposed what he thought of as the Christian devaluation of the body and the bodily instincts. The mind, as an entity distinct from the body, was a clever invention—in other words, a lie (1966, WII, 1157). It doesn't exist. Because the invented mind used to be taken as a proof of man's divine origin, one could only hope to reach human perfection by retracting, tortoise-like, one's senses into oneself, relinquishing all commerce with earthly things, discarding one's mortal shell, and thus retaining only what was essential to our humanity: pure spirit. For Nietzsche, however, "pure spirit" was "pure folly", and consciousness, in general, a "symptom of imperfection" (1966, WII, 1174). Nietzsche's will to power, which is the essence of all life, and in fact the essence of all being, is preconscious and non-rational, although it has its own, superior reason. One characteristic of the overhuman is that he knows himself to be "entirely body and nothing else" (1966, WII, 300).

Transhumanism "stresses the moral urgency of saving lives", which makes anti-ageing medicine "a key transhumanist priority" (Bostrom 2005b, 9). The indefinite extension of our life spans is believed to be an obvious good. Nobody wants to die, death is an evil, and life generally (though not necessarily under any circumstances) a good. Hence, if we could achieve personal immortality, we should not hesitate, but seize it. For Nietzsche, however, the promise of personal immortality is nothing but a "big lie" (1966, WII, 1205). Not so much because he thought it was impossible for us to ever become immortal, but rather because he believed that most of us are far too insignificant and worthless to deserve immortality. Promising immortality (or indefinite life extension) to everybody only boosts the widespread delusion that the world revolves around every single one of us, whereas in fact, most of us should never have been born in the first place. Most people actually die too late, not too early, because they have never learnt to live (1966, WII, 333). "'Immortality', granted to every Peter and Paul, has been the biggest, most vicious attack against noble humanity to date" (1966, WII, 1205). The promise of personal immortality pretends that we are all equal. It denies difference and rank. Moreover, it is based on an erroneous reification (Versubstanzialisierung) and atomisation of the individual self. The ego is wrongly differentiated from the non-ego, which are in fact inseparable in the eternal process of becoming (1966, WIII, 612). By wishing for personal immortality, I cut myself off from this process, believe myself to be more important than the rest of the world, which, for all I care, may perish if only I will be safe (1966, WI, 753). That is not an affirmation of power, but on the contrary, an indication of impotence. That is why, just like the human, the self or the "I is something that needs to be overcome" (1966, WII, 303). Instead of doing everything to escape death, we ought to practise the art of going at the right time and celebrate our dying as something that we freely embrace (1966, WII, 334), in order to plunge again into the great "ocean of becoming" (1966, WI, 1193), in which we belong. The overhuman understands how to live and how to die. The transhumanist, in Nietzsche's view, understands neither.

Yet if the overhuman is not an improved version of the human, what is he? There are of course statements in *Thus Spoke Zarathustra*, especially in the first sections, that sound as if Nietzsche was indeed advocating the transformation of the human into some kind of post-human. “Man is something that needs to be overcome” (1966, WII, 279). The overhuman is “the meaning of the earth” (1966, WII, 280), and man merely a “rope tied between animal and overhuman” (1966, WII, 281). But Nietzsche has no clear concept of the overhuman and produces at best vague intimations of what he has in mind (Shapiro 1980, 171). There is a chance that his overhuman is merely an ironic device, never meant to be taken seriously as an ideal human (Ansell-Pearson 1992, 310). After all, we shouldn’t forget that the overhuman was preached by Zarathustra, not Nietzsche himself, and may well be understood as a provisional concept in the ongoing movement of understanding (Lampert 1987, 258), as one possible perspective on the way things are, but not necessarily a true one, let alone *the* true one (Ansell-Pearson 1992, 314). Nietzsche himself warned of misunderstanding the overhuman as some kind of higher human. Zarathustra, he reminds us, is the destroyer of all morality, not half-saint, half-genius, not an idealist type of higher human, not a Parsifal, but a Borgia (1966, WII, 1101). He is mainly characterised by contempt: of personal happiness and of reason (TSZ, WII, 280). The overhuman is not thought of as an exemplar of a future human or post-human race, but as the “exceptional human” (Ausnahme-Mensch) (1966, WII, 1155), and there have always been such exceptional humans who were “in relation to the whole of humanity a kind of overhuman” (1966, WII, 1166). Even though Nietzsche sometimes talks as if a whole race of overhumans were possible, the overhuman can in fact only exist in the singular, that is, set apart from others. Overhuman is who is strong enough to take reality as it is, in all its fearfulness (EC, WII, 1156), with all its pain and suffering, who does not want anything different, to the point that he would welcome the opportunity to live it all again, just as it was. The eternal recurrence of the same, the idea of which is the true centre of the *Zarathustra*, is counter to the dynamic optimism that characterises transhumanist thought, and its non-selective affirmation by the overhuman counter to transhumanism’s morally toned selectivity.

All this makes it very unlikely that Nietzsche would, as Sorgner (2009, 34) claims, "have been in favour of genetic engineering" or indeed the transhumanist movement as a whole.

As we have seen in this and the last chapter, the transhumanist worldview is supported by two different and diametrically opposed conceptions of nature, namely: (a) a nature that limits us as organic-corporeal beings, confines us in particular forms of life, and curtails our autonomy; and (b) a nature that expresses itself in our reasoning faculties and our will, is our real essence, cannot but rebel against that other, confining nature, and whose final goal is complete dislimitation, the attainment of perfect autonomy. This opposition reveals a dualistic, very un-Nietzschean and almost Manichean idea of the human, according to which the body is to be understood as our "evil" nature, which we must seek to overcome, and the mind (and hence the will, which is informed by the mind) as our true, "good" nature, which we need to protect and nourish.

In contrast, those who have expressed serious doubts about the possibility and desirability of the proposed radical enhancement of the human, the so-called bioconservative critics such as Leon Kass, Michael Sandel, or Francis Fukuyama, can be recognised by their refusal to accept this basic dichotomisation of human nature (and they are in *this* respect closer to Nietzsche than to the transhumanists). Generally, bioconservatives are not particularly worried about the fact that we are limited beings. On the contrary, they are inclined to see our various limitations as a good thing: as that which gives us an identity, creates values, and opens up possibilities. If we are limited in all sorts of ways, then those limitations are exactly what makes us what we are, not only in terms of our weaknesses, but also in terms of our strengths. All the good that we can ever experience, we can only experience in the context of such limitations. Accordingly, bioconservatives also have an attitude to the body that differs considerably from that of the transhumanists. The human body is part of human nature, and it is precisely its fragility and vulnerability (so abhorred by those who set their hopes on technology to create better humans) that is deemed both intrinsically valuable (for instance, because it embeds us in a human community) and simply an integral part of human existence, which cannot be removed without thereby changing various other aspects of our being that we hold dear and that we neither want to lose, nor should lose. For the bioconservatives, there is no dichotomy between nature as limitation and nature as the (resistance-allowing, dislimitation-seeking) essential core of one's being. Rather, our specific way of being limited is wholly and undividedly our nature, for good and for bad, which is to say that we can do what we can do also because of all the things that we can*not* do. Seen from this perspective, the comprehensive control of our own existence that seems to be the goal of the whole transhumanist enterprise is not at all desirable, not the least because the reason why we value many aspects of our existence is precisely that they have *fallen* to us, that we cannot control them, that they elude our power. Love, happiness, friendship, all kinds of experiences, life itself, all this falls to us. We find ourselves in it, and this is an integral part of why we value it.[4]

This appreciation of our given nature does not commit the bioconservative to any kind of human nature essentialism, as has been claimed by, for instance, John Banja (2011). While I agree in principle with both Banja's moral relativist claim that there are no absolute moral categories and with his anti-essentialist position, it seems to me that his attack on moral conservative opponents of radical human enhancements fails on two counts: first, because it greatly exaggerates the extent to which moral conservatives are committed to essentialism, and second, because those who promote and defend radical human enhancement can be attacked with the very same arguments that Banja uses to discredit their moral opponents. It is true that moral conservatives sometimes present their arguments in a fashion that may easily be understood to indicate some essentialist commitment. Moral conservatives such as Francis Fukuyama, Leon Kass, Jürgen Habermas, or Michael Sandel clearly want to save human nature from the grasp of biotechnological renovations, fearing that by trying to improve on human nature, we might lose just as much as, or even more than, we will gain. Kass seeks to preserve "aspects of our given humanity that are rightly dear to us" (2003, 20), Sandel "the gifted dimension of human experience" (2007, 89), and Fukuyama the "full range of our complex, evolved nature against attempts at self-modification" (2002, 172) in order to maintain the basis of human dignity and rights, while Habermas is worried that "the instrumentalization of human nature changes the ethical self-understanding of the species in such a way that we may no longer see ourselves as ethically free and morally equal beings guided by norms and reasons" (2003, 40). All this talk about human nature certainly sounds as if there was some idea of a (normatively significant) human essence behind it, but this is in fact not the case. Note how Fukuyama defines the term "human nature": "[H]uman nature is the sum of the behaviour and characteristics that are typical of the human species, arising from genetic rather than environmental factors" (2002, 130). Hence, human nature, for Fukuyama, is more a prototype than an essence. When he occasionally uses the term "essence", he understands and explains it in terms of meaning (instead of the other way around, as Banja assumes). In other words, what he has in mind is what John Locke in his *Essay Concerning Human Understanding* called the nominal essence as opposed to the real essence of a thing. We have already discussed this in the third chapter: the nominal essence is a loosely connected cluster of recognisable properties that make up the abstract idea to which we attach a general name such as "human". And this is in fact all that moral conservatives need to support their arguments, which are without exception what Bill McKibben once called "arguments from meaning" (2003, 45). In order to make sense of these arguments, no "super-realities" (Banja 2011, 3) or "necessary, universally distributed, ultimately defining" properties (Banja 2011, 8) need to be posited. Neither do they rely on the "truth" of a particular account of human nature. As we have seen, the principal question that those thinkers raise, namely what it means to be human, is not a question that has a factual answer (nor do they suppose it does), but one that invites a critical examination of the goals that we pursue, the values that we endorse and promote, and the self-understanding that governs our actions. Of course, when Kass talks about human dignity and what it requires or Sandel about humility, they both commend a certain personal view of what makes human life worthwhile, what is important in life and what is less so (Hauskeller 2011b). And of course these answers are "constructs" and "local" (Banja 2011, 15), but that is what views about how we should lead our lives inevitably are. We all have our own mythologies to guide us. There are no definite answers to ethical questions, so we shouldn't expect moral conservatives to produce such answers. And they are not trying to. Instead, they offer an alternative to the seemingly common-sensical arguments favoured by transhumanists and in general all those who think that we should do all we can to develop technologies that allow us to "make better people" (Harris 2007).

When Banja criticises moral conservatives for their allegedly "dogmatic, certain, and absolute" argumentative style, which is ideally suited to "generate support for their favored brace of moral beliefs" (2011, 20), he conveniently forgets that proponents of radical human enhancement do not seem to be any less certain or dogmatic when they present their views. Thus, John Harris claims not only that there is "a clear imperative to make the world a better place" (2007, 4) and that we have a moral duty to enhance, but also that it is our duty to pursue and participate in research which helps develop enhancement technologies. According to Harris, this is not a matter of opinion, but simply a matter of fact. There is no room for doubt. Likewise, many proponents of human enhancement are united in the view that death is "the greatest evil" (More 1996) and therefore finding a cure against ageing is nothing less than an "urgent, screaming moral imperative" (Bostrom 2005, 277). There doesn't seem to be much uncertainty here. The same holds for those who, instead of proclaiming a moral duty to pursue certain kinds of enhancement, simply want us to remain free to make our own choices, including the choice to enhance ourselves or our children. They, too, don't seem to have any doubt that the liberty to enhance and to "seize control of our evolutionary future" (Stock 2003, 2) is something worth defending against common worries. The truth is that neither moral conservatives nor—how shall we call them?—moral progressivists tend to be moral relativists. They all believe that certain things are right (or at least more desirable) and others wrong (or less desirable); it's just that they differ about what these things are.

Furthermore, as we have seen, many of the arguments proposed in favour of radical human enhancement tacitly or openly appeal to a particular *essentialist* conception of human nature. When Gregory Stock (2003, 2) talks about our Promethean nature that compels us to continue "stealing fire from the Gods", about the "characteristically human" trait of not accepting any fixed boundaries, which makes biological enhancement inevitable, then he clearly appeals to a supposed essence of humanity that not only justifies the project of "redesigning" ourselves by means of technology, but also leaves us no choice but to go along with it. Others are even more explicit, for instance, when Julian Savulescu, in his defence of performance-enhancing drugs in sport, curtly declares that to "choose to be better is to be human" (2004, 670) and therefore drugs in sport should be allowed. According to Savulescu, what makes us human is our ability to use our rationality for self-improvement. The rational animal is here reinterpreted as an essentially self-enhancing animal. The implication is that, precisely for that reason, we should be allowed (and indeed encouraged) to use all available means to achieve the goal that is intrinsic to our existence (and particularly evident in sport). This normatively laden, and no doubt essentialist, understanding of the human is rather common among transhumanists and other proponents of radical human enhancement. To give one last example, Nick Bostrom declares human nature to be "dynamic, partially human-made, and improvable" (2005b, 213) and infers from this that human dignity must consist in "what we are and what we have the potential to become". The potential to become something other than what we are is thus not only what makes us human, but also what gives us that special worth on which all our moral rights ultimately depend. Hence, to turn our backs on this potential would both violate our nature and compromise our dignity. Can an argument get any more essentialist than this?

But even if this were not so, even if comprehensive, all-encompassing control (i.e. a complete dislimitation and natureless existence) were indeed desirable, it could be the case that the whole plan must still fail, simply because it is self-contradictory, as the British writer and philosopher C.S. Lewis (1955) argued many decades ago against the "conditioners" of his own time. In his book *The Abolition of Man*, which is as current today as it was 70 years ago, he analyses the then widely used expression of "Man's conquest of Nature", wondering in what sense exactly we can say that "man" has conquered, or gained more control over, nature. He comes to the conclusion that, first of all, the power that we gain over nature through the use of new technologies is not really *our* power at all. Rather, the power belongs to the technology itself, which we use, but which we can also lose at any time. We can temporarily control more things, but we also become more dependent. The power that we seem to have gained is in fact only borrowed. It is a power by proxy, and as such can very quickly turn into an even greater powerlessness if the actual source of the power suddenly refuses to collaborate. One single error in the system renders us helpless. We can perhaps steal fire from the Gods, but the fire still belongs to them, and they can call it back any time.

Second, the power that we possess as humans through the technology available to us is never possessed by *all* humans. That power lies in fact always in the hands of *some* people, who can then use that very same power against other people. The power that "we" have is thus also a power that we are all, at least potentially, victims of. The powerful bomb that I develop can always end up being used against me and thus destroy me. Every increase in power also increases our vulnerability.

Third, the idea of total power and control, achieved through science, is self-contradictory. To be consistent, the conditioner also needs to jettison the values and goals that direct his own actions and the use he makes of science and technology, because they, too, can no longer be taken as a given. They, too, must be controlled and become the product of a deliberate act of design: "[I]t is the function of the Conditioners to control, not to obey them. They know how to produce conscience and decide what kind of conscience they will produce" (Lewis 1955, 74). Yet on what basis should they decide which values they want to follow? Science itself cannot provide what they need, for, as Karl Jaspers (1971, 10) has rightly pointed out, it "can give no answer to the question of its own meaning. The existence of science rests upon impulses for which there is no scientific proof that they are true and legitimate". Without such a basis, every decision becomes arbitrary and the product of a mere whim. This means, however, that all our decisions are now entirely accidental and cease to be *our* decisions in any meaningful way. Decisions are being made without reason, which means that they are in fact being made for us. Once our control is complete, we are nothing more than the pawn of our whims, so paradoxically, as Lewis points out, nature, as it is now freed from all values, controls the conditioners and through them all humankind. "Man's conquest of Nature turns out, in the moment of its consummation, to be Nature's conquest of Man." (Lewis 1955, 80).

Fourth, we reduce ourselves to nature by turning ourselves into something controllable. According to Lewis, we call "nature" anything and everything that can in principle be controlled, so the price that we pay for making ourselves (our emotions, our conscience) an object of control is that we must now see and treat ourselves as just another piece of nature.

By completely controlling nature, including our own, we give even more room to nature. Everything has become nature, so by conquering it, it has conquered us. Lewis sees this as a magician's bargain, through which we sacrifice our soul, that is, our self, to gain power. But as in any proper magician's bargain, this power does then not really belong to us at all, but to the one to whom we have sold our soul: "[I]f man chooses to treat himself as raw material, raw material he will be: not raw material to be manipulated, as he fondly imagined, by himself, but by mere appetite, that is, mere Nature, in the person of his dehumanized Conditioners" (Lewis 1955, 80).

To see how right Lewis was in his assessment, one only has to look at the way new enhancement technologies are actually being used and, more importantly, what kind of use people seem to be interested in. A nice example is the molecule oxytocin, which is a hormone found in mammals that also functions as a neurotransmitter and that has gained celebrity status as "love hormone" or "cuddle hormone". Allegedly,[5] it enhances our social competence, makes us nicer and more considerate, more sociable and sympathetic to the plight of others, generally more trusting and, at the same time, more confident (Zak 2012). It is even supposed to boost the male sexual drive. Naturally, all of this makes it seem quite appealing, which hasn't been lost on the pharmaceutical industry, so we can now buy oxytocin in form of sprays (to be applied to one's clothing), which by some is being celebrated as a major achievement, as an important step towards the urgently needed moral improvement of humanity. However, the marketing strategy of those sprays conveys a very different message and shows clearly what is really going on here. The aptly named *Liquid Trust Enhanced*, for instance, which is produced by Vero Labs and is advertised as "trust in a bottle", is openly marketed as an extremely efficient means to manipulate other people and to get them to do what you want them to.[6] Under the slogan "trust is power", the company proudly lists all the good uses to which the spray can be put and all the advantages that using it can bring you. If it is your job to sell stuff, then *Liquid Trust Enhanced* will help you to sell even more of it, because when you spray yourself with it, people will feel inclined to trust you: "Close the deal! Sell more of your products and services." But even if that doesn't work out, you can still be certain to get a pay rise because your boss will, without needing a reason, be impressed by you and will put his trust in you if you beguile him with your oxytocin smell: "Give your career a powerful boost! Ask for a raise – And GET IT!" There are also advantages for your private life because the spray greatly facilitates building new relationships, especially of a sexual nature: "Meet more women because they trust you more." In short, this wonderful product is really "the ultimate sales tool", "the ultimate tool for relationships", and "the ultimate career builder".

The example shows how the enhanced human of the transhumanist imagination becomes better at manipulating the world around him (including other people), but also, and for the same reason, much more vulnerable to the manipulation of others. The more extensive the control is we have over the world, the more extensive is the control the world has over us. Thus, the abovementioned *unnature* of the enhanced human, the attainment of which is the goal of the whole radical human enhancement project, not only goes along with a complete naturalisation of the human, but also ultimately puts an end to our existence not just as humans, but even as the kind of human or post-human that transhumanists would like to see us become.

It is the nature of power that we all cannot have it. The more one has, the less others have, and if one individual has *all* the power (and only then can they be completely autonomous), then nobody else can have any power. Naturally, this is a fact that most transhumanists are reluctant to admit, perhaps even to themselves. Some, however, are quite willing to spell out the rather unpleasant implications of their commitment to transhumanist ideals. I am thinking mainly of the American journalist Zoltan Istvan, who in 2014, founded the Transhumanist Party and is currently touring the USA with his coffin-shaped Immortality Bus to convince people that he should become the next American president. Three years ago, Istvan published a novel that outlines his understanding of the transhumanist philosophy. *The Transhumanist Wager* is a terrible novel by all measures, full of clumsy descriptions, wooden characters, and even woodener dialogues, but it is still worth reading because it gives us some idea of why Fukuyama (2009) may not have been entirely wrong when he nominated transhumanism as one of "the world's most dangerous ideas". Istvan's novel makes transhumanism look like the brainchild of

Ayn Rand and Adolf Hitler. It is basically a transhumanist blend of *Atlas Shrugged* and *Mein Kampf*. It reads like a dystopian novel, but is apparently meant to be eutopian, which makes it even scarier. The novel starts with the three laws of transhumanism (Istvan 2013, 4):

1. A transhumanist must safeguard one's own existence above all else.
2. A transhumanist must strive to achieve omnipotence as expediently as possible—so long as one's actions do not conflict with the First Law.
3. A transhumanist must safeguard value in the universe—so long as one's actions do not conflict with the First and the Second Law.

After that, we don't hear much about the third law and what "value in the universe" actually consists in. However, we do hear a lot about the importance of self-preservation at all costs and the power that the "elite transhuman champion", or "omnipotender", rightfully wields over other people. The "omnipotender is an unyielding individual whose central aim is to contend for as much power and advancement as he could achieve, and whose immediate goal is to transcend his human biological limitations in order to reach a permanent sentience" (33). Other people, especially other transhumanists, may be usefully employed to help the omnipotender to accomplish this goal, but ultimately, he knows that he is "fundamentally alone in the universe" (12). The novel's transhumanist hero, Jethro Knights (a thinly disguised incarnation of the author), is not much bothered by this. He has little interest in other people: Jethro "rarely listened to people. Or noticed them at all. Even if he looked a person directly in the eye, he often failed to recognise anything of utility. Jethro perceived their presence, the space they took up, the resources they used on this planet. His brain interpreted the matter and energy they possessed, but unless there was potential for something useful to him, he may as well have been looking at a rock, or a weed, or a broken, outmoded piece of furniture in a junkyard. Jethro only took notice of values, not people" (12).

The novel describes Jethro's rise from a social underdog and failed philosophy student to a charismatic leader of the transhumanist party and eventually the undisputed ruler of a new thoroughly transhumanised earth, a global state called Transhumania, in which every single citizen has no other interest but to work towards achieving the transhumanist goal of conquering death and generally all biological limitations. They are given no choice. Those who have no interest in that goal, or simply nothing to contribute, are killed or left to die. According to the transhumanist philosophy advocated in the book, which Jethro calls Teleological Egocentric Functionalism, they do not deserve to live. Nobody that "stands in the way of our transhuman way of life" does. Before he takes over the world by attacking all resisting nations with superweapons developed by transhumanist scientists, he proudly announces to the world that he will shy away from nothing to "build a brave new reality and vision for the world": "The morality of Transhuman Citizen is defined and decided by the amount of time we have left to live. Not by democracy, decency, altruism, kindness, or notions of humanity and mammalian love" (85). All that has to be thrown overboard, *not* for the sake of the greater good of humankind, but to remove all obstacles that might hinder the omnipotender to claim his "birthright" of acquiring as much power as possible. This is, after all, "the essence of evolution" (88), and therefore "the most natural outlook on reality" (273). To that end, it is necessary to divert all resources "to the genuinely gifted and qualified. To the achievers of society – the ones who pay your bills by their innovation, genius, and hard work. They will find the best way to the future. Not the losers of the world, or the mediocre, or the downtrodden, or the fearful. They will only drag us down, like they already have" (128). The message is loud and clear: no room in the inn for "welfare-collecting non-producers" (200) and "the slackers and freeloaders around the world who don't want to work" (285). And no pity or mercy either: "We will kill you if we have to. If needed, we will kill every one of you, down to the last enemy of transhumanism on this planet. We will eliminate you into the void of the universe with no remorse, with the same cold morality a machine would use. We are through playing by your rules and on your terms" (202). But even some of those who are quite willing to play the game by transhumanist rules may have to be eliminated if found wanting, that is, not useful enough. Transhumanist scientists have already developed an algorithm to determine a person's net value: "Any individual who ultimately hampers the optimum transhuman trajectory of civilisation should be eliminated. The Humanicide Formula addresses this issue directly. It determines whether an individual should live or die based on an algorithm measuring transhuman productivity in terms of that individual's remaining life hours, their resource consumption in a finite system, and their past, present, and potential future contributions" (215).

I don't think there is much to comment here. Those statements speak for themselves. I realise that the book is a novel, but I cannot detect any irony, any attempt of the author to distance himself from the views of the novel's main protagonist. So I am assuming that what Jethro Knights says is more or less what Zoltan Istvan, the presidential candidate of the American Transhumanist Party, thinks. It is a chilling amalgam of technophilia, ethical egoism, social Darwinism, anti-consumerism, anti-egalitarianism, anti-welfarism, and anti-religionism. And although Istvan's views may not be representative of the movement as a whole, I am wondering whether they do not reveal the secret soul of transhumanism, that which is usually left unsaid by its main proponents because they are too cautious or perhaps simply too decent to allow themselves to think that way. But the omnipotender lives and breathes in all of them. If so, then perhaps More is right after all to claim Nietzsche as an inspiration for transhumanism: it is the Nietzsche who expressed his contempt for the "morality of the herd", who did not believe in equality but in the supreme right of the strong, the admirer of the human predator, the splendid blond beast who takes what he needs and redefines what is good and bad. As Jethro Knights does, the transhumanist hero, herald of a new age: "We live according to what we believe we are becoming: we call it the futurization of all values. (…) Our interpretation of values taught us that evolution and its ascent of technology do not operate off democratic principles, but off principles of might, off principles of survival. (…) Nothing and no one is equal. (…) We should embrace that and act on it. Throughout history, it's the reason the strong became stronger and the weak became weaker, until the weak were no more."

Notes

1. For instance, to name but a few, Max More, Nick Bostrom, Aubrey de Grey, and John Harris.

2. Cf., for instance, Harrington (1969, 182), or James Stacey Taylor (2009, 109), who regards ageing and death as "biological constraints" of autonomy.
3. Savulescu's principle is not usually thought to be so far-reaching in scope. However, the postulation of a parental moral obligation to select the best *possible* children suggests that the use of available enhancement technologies to optimize our offspring is also obligatory.
4. For a more detailed justification of this claim, cf. Hauskeller (2011b).
5. I am here referring to *popular* representations of the hormone, not to its *actual* effects. For a more nuanced and evidence-based discussion of what oxytocin can and cannot do, see Wudarzcyk et al. (2013).
6. https://www.verolabs.com/Default.asp?affl=sas

Bibliography

Ansell-Pearson, Keith (1992) "Who is the Ubermensch? Time, Truth, and Woman in Nietzsche", *Journal of the History of Ideas* 53/2: 309–331.

Banja, John (2011) "Virtue essentialism, prototypes, and the moral conservative opposition to enhancement: A neuroethical critique", *AJOB Neuroscience* 2/2: 31–38.

Bostrom, Nick (2005) "The Fable of the Dragon Tyrant," *Journal of Medical Ethics* 31: 273–277.

——— (2005a) "A History of Transhumanist Thought", *Journal of Evolution and Technology* 14/1: 1–25.

——— (2005b) "Transhumanist Values", *Review of Contemporary Philosophy* 4.

Fukuyama, Francis (2002) *Our Posthuman Future*, New York: Farrar, Straus & Giroux.

——— (2009) Transhumanism, *Foreign Policy*, http://foreignpolicy.com/2009/10/23/transhumanism/

Habermas, Jürgen (2003) *The Future of Human Nature*, Cambridge: Polity Press.

Harrington, Alan (1969) *The Immortalist*, New York: Random House.

Harris, John (2007) *Enhancing Evolution. The Ethical Case for Making Better People*, Princeton: Princeton University Press.

Hauskeller, Michael (2009) "Prometheus Unbound. Transhumanist Arguments from (Human) Nature", *Ethical Perspectives* 16/1: 3–20.

——— (2011a) "Forever Young? Life Extension and the Ageing Mind", *Ethical Perspectives* 18/3: 385–406.

——— (2011b) "Human Enhancement and the Giftedness of Life", *Philosophical Papers* 40/1: 55–79.
Istvan, Zoltan (2013) *The Transhumanist Wager*, Futurity Imagine Media.
Jaspers, Karl (1971) *Philosophy of Existence*, Philadelphia.
Kass, Leon (2003) "Ageless bodies, happy souls", *New Atlantis* 1/1: 9–28.
Lampert, Laurence (1987) *Nietzsche's Teaching. An Interpretation of Thus Spoke Zarathustra*, New Haven: Yale University Press.
Lewis, C.S. (1955) *The Abolition of Man* [1943], New York: Macmillan.
McKibben, Bill (2003) *Enough. Genetic engineering and the end of human nature*, London: Bloomsbury.
More, Max (1996) "Transhumanism. Towards a Futurist Philosophy", www.maxmore.com/transhum.htm.
——— (2009) "True Transhumanism", *Global Spiral*, February.
——— (2010) "The Overhuman in the Transhuman", *Journal of Evolution and Technology* 21/1: 1–4.
Nietzsche, Friedrich (1966) *Werke in drei Bänden* (W). Ed. Karl Schlechta, Munich: Hanser Verlag.
Pico della Mirandola (1985) *On the Dignity of Man, On Being and the One, Heptaplus*, New York: Macmillan.
Sandel, Michael (2007) *The Case Against Perfection. Ethics in the Age of Genetic Engineering*, Cambridge, MA: Harvard University Press.
Savulescu, Julian (2001) "Procreative Beneficence: why we should select the best children", *Bioethics* 15(5/6): 413–426.
Savulescu, Julian, Bennett Foddy, and Megan Clayton (2004) "Why We Should Allow Performance Enhancing Drugs in Sport", *British Journal of Sports Medicine* 38: 666–670.
Shapiro, Gary (1980) "The Rhetoric of Nietzsche's Zarathustra", *Boundary* 2/8: 165–189.
Sorgner, Stefan (2009) "Nietzsche, the Overhuman, and Transhumanism", *Journal of Evolution and Technology* 20/1: 29–42.
Stock, Gregory (2003) *Redesigning Humans*, London: Profile Books.
Taylor, James Stacey (2009) *Practical Autonomy and Bioethics*, New York/London: Routledge.
Wudarczyk, Olga, Brian D. Earp, Adam Guastella, and Julian Savulescu (2013) "Could intranasal oxytocin be used to enhance relationships? Research imperatives, clinical policy, and ethical considerations", *Current Opinion in Psychiatry* 26/5: 474–484.
Zak, Paul J. (2012) *The Moral Molecule. The new science of what makes us good or evil*, London: Transworld Publishers.

6
Fixing the Animal

In the second chapter of this book, I claimed that transhumanists, in contrast to critical post-humanists such as Haraway (1985),[1] Hayles (1999), or Wolfe (2010), have a decidedly *humanist* understanding of the world and our place in it. Transhumanists, I said, do not doubt that humans are special, that reason sets us apart from the rest of nature, and that we all carry the potential in us to ascend the heavens and to be and live like Gods. It is now time to qualify that statement. The term "humanist" does not have an exact definition. Just as there are many ways of being human, there are also many ways of being human*ist* (and, accordingly, many ways of being *post*humanist). Max More, in his introductory chapter to *The Transhumanist Reader* (More and More 2013), is happy to acknowledge that transhumanism has roots in Enlightenment humanism, but for him, that seems to mean not much more than that it is forward-looking, secular, and rational. What is humanist in transhumanism is the "emphasis on progress (its possibility and desirability, not its inevitability), on taking personal charge of creating better futures rather than hoping or praying for them to be brought about by supernatural forces, on reason, technology, scientific method, and human creativity rather than faith" (More 2013a, 4). There is little to argue with here, at least for me: the

M. Hauskeller, *Mythologies of Transhumanism*,
DOI 10.1007/978-3-319-39741-2_6

intention to create better futures for ourselves and others and the reliance on reason rather than faith is pretty much the default position in Western societies today. Most of us would not want it any other way.

Yet there may be other forms of humanism in transhumanism that are less innocuous. So what I am going to do in this chapter is determine in what way transhumanists are humanists and in what way they are *not*. To do this, let us first briefly focus once more on the role that the idea of autonomy plays in the transhumanist worldview, and then, in more detail, on the transhumanist attitude towards non-human animals, which will make up the bulk of this chapter.

According to Tamar Sharon (2014, 3), transhumanism—or, as she refers to it, "liberal posthumanism"—is "grounded in the humanist narrative of the human as an autonomous, unique and fixed entity that is separate from its environment in a distinct way". Yet even if that is *the* humanist narrative, it does not quite capture the *trans*humanist narrative, which is rather different. It is true that a transhumanist would typically be someone who believes that individual autonomy is a very important good, perhaps even the highest good, perhaps even the only (intrinsic) good (although the attainment of pleasure—or the absence of pain—may be even more important for some transhumanists). Accordingly, whatever increases individual autonomy is good and ought to be sought and supported, and whatever diminishes or in any way limits or compromises individual autonomy is bad and ought to be avoided and fought. Human enhancement technologies promise to increase individual autonomy and therefore deserve our support. This, I take to be a key tenet of the transhumanist mythology. If you don't think that individual autonomy is particularly important, that there are other things that might be more important, or that people should not be given access to enhancement technologies to improve themselves and actively work on their own self-creation, then you are not a transhumanist. However, what a transhumanist does not have to believe is that the subject already *is* autonomous. It is precisely because the transhumanist is keenly aware that we are *not* autonomous, not self-contained, and not separated from our environment, at least not sufficiently so, that they set their hopes on enhancement technologies, the soon-to-be-expected fusion of human and machine, and ultimately, the complete digitalisation of our existence and identity. Autonomy, for the transhumanist, is a value, not a fact. It may well be a humanist value, but upholding it does not betray any erroneous assumptions about the nature of the human subject, as Sharon believes.

Perhaps more importantly, the autonomy that is sought by the transhumanist does not require a commitment to any kind of fixed, unchanging essence. Usually, transhumanists are quite relaxed about personal identity issues raised by their opponents. Will it matter if the post-human that we eventually transform into is so different from us that it would be difficult to regard us both as the same person? Not really, says the transhumanist, certainly not if we *value* self-transformation, as transhumanists are wont to do (More 1995). If radical enhancement turns us into something radically different, so be it. Change is good, radical change even better. Personal identity is overrated anyway. James Hughes wants to discard the notion of the self and imagines future post-human societies as "post-personal identity societies" (Hughes 2013). Will it matter if the self that exists after I have managed to upload my mind to a computer is not really me, but only in all relevant respects like me? Not really, says the transhumanist. An exact copy of me is as good as me because it *is*, for all intents and purposes, me. The sameness of the *pattern* is what counts (Kurzweil 2005). What if, after the singularity, we do not merge with machines, but are actually being replaced by them? Fine, says the transhumanist. We are not partial, and more than happy to regard machines as our legitimate heirs, our "mind children" (Moravec 1988; Kurzweil 2005, 260). What if, after we have managed to radically extend our life spans, we are required to periodically restart our lives (and forget our previous ones) to avoid getting bored with ourselves and prevent mental ageing? No problem. The main thing is that we are free to live any life we fancy, and that we are indeed free to be anyone we want to be. Far from clinging to a fixed self, the transhumanist is hell-bent to get rid of it. All determination is a limitation, and all limitations are bad. What Sharon says about radical (and that is, genuinely post-humanist) post-humanism, namely that it shrinks from the "terror of fixed and unified identity" and that it claims "the right to difference, variation and metamorphosis" (Sharon 2014, 151), one could just as well say about transhumanism. Moreover, if in our effort to overcome that fixed self, we will eventually merge with everyone else and become part of a group mind, or even merge with a thoroughly spiritualised universe so that any real separation between ourselves and our environment, between self and other, is effectively suspended, then this is just as it should be, because as long as there is an Other (which necessarily limits our existence), we cannot be truly autonomous and free. So, in sum, I don't think that, on closer inspection, it is true at all that "liberal post-humanism [read: transhumanism] retains an account of supplemental prostheticity of encounters with technology all the same: some initial, unified self remains intact and essentially unpenetrated by new technologies" (Sharon 2014, 98). In fact, most transhumanists cannot wait to be thoroughly penetrated and transformed by new technologies, both in body and in mind. So if by humanism we mean a commitment to an unchanging human essence, then transhumanists are not humanists.

Nor are they humanists in the sense that they believe that humans are the only beings worthy of moral consideration, the only ones that have true moral standing. They do not subscribe to the kind of *ethical* humanism that characterised the philosophy of, for instance, Thomas Aquinas or Immanuel Kant. Instead, most transhumanists follow the utilitarian tradition, which emphasises the ability to suffer as a normatively relevant common ground between humans and animals. Since animals are sentient creatures, they do deserve at least *some* moral recognition. Thus, David Pearce, author of the transhumanist manifesto *The Hedonistic Imperative*, in which he advocates the biotechnological abolition of *all* suffering, including that of non-human animals (Pearce 1995), states that from "a notional God's-eye perspective, I'd argue that morally we should care just as much about the abuse of functionally equivalent non-human animals as we do about members of our own species – about the abuse and killing of a pig as we do about the abuse or killing of a human toddler" (Pearce 2007). Along the same lines, the *Transhumanist Declaration*, crafted in 1998 by Nick Bostrom, David Pearce, Max More, and others, and later officially adopted by the world transhumanist association Humanity Plus, explicitly commits transhumanists to the advocacy of "the well-being of all sentience, including humans, non-human animals, and any future artificial intellects, modified life forms, or other intelligences to which technological and scientific advance may give rise" (Humanity Plus 1998). Other transhumanists emphasise the fact that at least some non-human animals qualify as (Lockean) persons, and demand that human-level legal rights be conferred to them (or in general to all "non-human persons", which of course also includes, or would include, intelligent, self-aware machines). Under the leadership of George Dvorsky, the Institute for Ethics and Emerging Technologies promotes a "Rights of Non-Human Persons" programme, which aims at defending "the rights of non-human persons to live in liberty, free from undue confinement, slavery, torture, experimentation, and the threat of unnatural death" (http://ieet.org/index.php/IEET/RNHP). Even in Zoltan Istvan's Transhumania (rather bizarrely, given the author's apparent scorn for all "useless" beings), the restaurants sell only "cruelty-free foods" (Istvan 2013, 197). So again, if by humanism we mean the belief that only humans count, that only human well-being matters morally and we have no obligations to non-human animals, then transhumanists are, as a group, not humanists.

However, transhumanists still believe that humans are special in the sense that they alone possess the ability to self-transform under the guidance of reason and in accordance with goals derived from a rational assessment of what matters in life and what is objectively good and worth having and being. We alone can make that assessment, and we alone can use our insight to redesign a suboptimal world, which includes redesigning our suboptimal selves, as well as those of others. This is our main obligation, our mission on earth. Non-human animals cannot take on that mission because even the most intelligent animals are stuck in the natural world, forever confined to the specific bodies and minds that they have been given by nature, condemned to accept their various inabilities: their comparative lack of understanding, the shortness of their lives, the inevitability of their deaths, because they have no choice in the matter. But *we* do. Our ability to reason makes a huge difference. While it does not make us autonomous, it gives us the *potential* to free ourselves from the confinements of nature. Just like non-human animals, we are currently still "slaves to our genes" and subject to "the tyranny of aging and death" (More 2013b, 450), but at least we have a good fighting chance to pull free of all that if we only put our mind (and its offshoot, science and technology) to it. To finally take up that fight in earnest is what transhumanists urge us to do. Thus, Max More, in a "Letter to Mother Nature", which starts with an acknowledgement of "the many wonderful qualities" that Nature has bestowed on us and ends in what looks more like an outright declaration of war on her, programmatically proclaims:

> We will take charge over our genetic programming and achieve mastery over our biological and neurological processes. We will fix all individual and species defects left over from evolution by natural selection. Not content with that, we will seek complete choice of our bodily form and function, refining and augmenting our physical and intellectual abilities beyond those of any human in history. We (…) will not limit our physical, intellectual, or emotional capacities by remaining purely biological organisms. While we pursue mastery of our own biochemistry, we will increasingly integrate our advancing technologies into our selves. (More 2013b)

This envisaged act of deliberate self-creation is what, in the transhumanist understanding, marks us as human. What we shall leave behind us by cutting all ties to Mother Nature is precisely everything that we have in common with non-human animals, with what is *not* distinctly human about us. What we shall leave behind, or "fix", is, in other words, the *animal* in us.[2] We kill the mother[3] so that we no longer have to be her sons and daughters, as all the other animals continue to be. Unless, of course, we do something about it. If we accept that our lives are poor and unsatisfactory, that we live the life of slaves (to our own biology) because we are ultimately (still) animals[4] (or perhaps transanimals[5]), then the lives of non-human animals must be judged the same. Severely limited in their possibilities as non-human animals are, which are even more limited than we are, their lives must be understood as even poorer than ours. While we at least have *some* degree of self-determination and potential for self-creation, they have none. Yet if we are fixable, then they may be too. And since transhumanism is a philosophy that officially subscribes to the view that all sentient creatures deserve moral consideration and, if needed, our help and support, as transhumanists, we have a duty to step in and fix not only ourselves, but also all other animals. Thus, according to James Hughes, we "have an obligation to children to provide them with education and secure homes so they can realize their abilities. We have an obligation to the mentally ill to provide them with treatments that return them to sanity. Alongside the provision of basic needs, education and a caring community, we also are increasingly able to offer technology as a means for people to reach their fullest potentials. (…) I think we have the same obligation to uplift 'disabled' animal citizens that we have to disabled human citizens" (Hughes 2004, 224).

The sentiment is noble perhaps, but also quite patronising. It is not compassion, but pity, that is being shown here, of the kind that we would resent if expressed towards us, because it always involves condescension, the presumption of superiority. Poor brutes, such lowly lives they have; let us take pity on them and lift them up to our own lofty heights! This is a far cry from what Donna Haraway (2008) describes as the meeting of species, which involves the practical recognition of the animal as a companion, as an equal, responsive, and active partner in the muddy dance of life. "I am a creature of the mud, not the sky", says Haraway (2008, 4). Not so the transhumanist, who decidedly leans towards the sky as his (and our) true home. Animals live in the mud, and children play in it. They know nothing of the sky. For Hughes, animals are like human children who are deficient because they have not developed their full potential yet. But at least children will one day grow up, nearer to the sky, whereas animals will never, or at least not without a little help from their friends, namely us. Animals are in a *permanent* childlike state, which here does not signify innocence, but immaturity and dependence. Only we can save them from the misfortune of a permanent childhood. And to add insult to injury, animals are also likened to the mentally ill and mentally disabled. Something significant is missing from their constitution, something that they ought to have but cannot acquire by themselves. We need to jump into the breach and help them, restore them to sanity.

The human is here figured as the better animal (precisely because we are less animal, or transanimal), just as the post-human is figured as the better human (because they are less animal even *more*). What the post-human is in comparison with us, we are in comparison with non-human animals. They are conceived as *pre*humans (in the same way that we are conceived, teleologically or at least trajectorially, as pre-posthumans). Consequently, we look after an animal's well-being by helping it to become something that is no longer animal. What is good for the animal (be it non-human or human) is that it disappears *as* an animal. The enhancement of the animal lies in its elimination; the only good animal is an ex-animal. This is, ultimately, what all proposals for animal enhancement suggest.

Transhumanist uplifting simply follows that tradition. What is different is merely the *kind* of elimination that is suggested.

As I have argued in *Better Humans?* (Hauskeller 2013), there is no such thing as human enhancement, understood as an enhancement of the human *as* a human, simply because being human is not something one can be good at, better at, or worse at. Humans can of course be better than others in all sorts of ways, depending on what we want them to be, but there is no such thing as a human that is *per se* better, that is, as a human. Similarly, the idea of making animals better only makes sense in the context of a particular set of purposes. Consequently, if an animal is to be made better, we need to know what exactly it is going to be better for or at, or in what *respect* it is going to be better. Normally, when we hear about enhanced animals, what people mean is animals that are improved in such a way that they better serve particular human interests and purposes. These can be agricultural, artistic, or social interests, or purposes related to any other area in which humans interact with animals. We can intend to make them better at hunting or as pets, for food production, or as companions. We can intend to improve their looks, their strength, or their health, or their entertainment value. We can want them to function more reliably as a model for certain diseases (e.g. oncomice), as generators for human pharmaceuticals ("pharming"), or as learning tools. This kind of (human interest–led) enhancement has been practised deliberately for a very long time, first through selective breeding and now increasingly also by means of gene transfer and other forms of direct genetic modification. Another new possibility of enhancing the animal is through cyborgisation, where a mechanical or electronic device is temporarily or permanently integrated into an animal's organism to render them more suitable to our purposes.

One interesting recent example is the roboroach produced and marketed by the small start-up company *Backyard Brains*. Roboroaches are cockroaches with an electronic device on their backs connected to the neurons in their antennas, which allows the user to control their movements with a mobile phone (http://www.kickstarter.com/projects/backyard-brains/the-roboroach-control-a-living-insect-from-your-sm): "When you send the command from your mobile phone, the backpack sends signals to the antenna, which causes the neurons to fire, which causes the roach to think there is a wall on one side. The result? The roach turns!" This will work only for a few minutes, though. The roaches quickly wise up and ignore the literally misguiding information. Hailed as "the first commercially available cyborg", the roboroach is claimed to be very useful to teach students "how our brains work". However, the main insights to be gained here are, first, that you can actually manipulate animals by messing about with their sense organs and brains and thereby tricking them into believing that things exist which in fact do not, and if that is possible, we may reasonably infer, then it may also be possible to trick the human brain into believing in the existence of things that do not exist. (But does that really come as a surprise? Didn't we know that already? And aren't there easier and more direct ways to demonstrate this?) The second, and by far more interesting, insight to be gained from the experiment is that it does not take the roaches long to realise that they are being tricked. How did they figure that out? Their senses tell them there is a wall, but after a short while, they know there is not, while (presumably) their senses still insist that there is. This is quite remarkable. So what have we really learned from this? How our brains work? Hardly. What we have learned is rather that even the brain of a cockroach—or more simply: a cockroach—is far more complex than our customary disregard for these creatures prompts us to believe. And also perhaps that reality has a way of reasserting itself. However, my guess is that most people will use the technique not to gain or generate that insight, or to learn anything at all, but rather because they are fascinated by the prospect of actually being able to remote-control a living being. This is certainly what the company's website suggests with the headline: "Control the movements of a live cockroach from your own mobile device!" To subject its will to our will, directly, without recourse to physical violence, that is an experience many may be quite willing to pay for. It is a kind of mind control, which is more complete than any other kind of control. It is a vision of ultimate power, which promises an excellent opportunity to satisfy what Michael Sandel (2007) has dubbed our drive to mastery. Thus, we are probably much more interested in the few minutes during which the cockroach actually does what we want it to do than we are in the power of agency that only all too soon allows it to free itself from our reign. That alone may be a good reason not to promote the widespread use of roboroaches.

The power of agency demonstrated by the roboroach is the power of the animal that refuses to be turned into the machine that we want it to be: something that can be controlled and reliably does our bidding. The machine is that which can be controlled; the animal is that which cannot. The purpose of most forms of animal enhancement is to suppress or eliminate everything in the animal that does not directly contribute to its ability to do what we want it to do. A certain function is assigned to it, perhaps based on already existing natural abilities or inclinations, and then we work on it until it has, ideally, become one with that function. Dog breeding is a good, more traditional example. For centuries, dogs have been bred and trained to fulfil highly specialised tasks relevant to particular human practices. Thus, various kinds of hunting dogs were created, each specialised in a particular type of game. Further specialisations followed. Bird dogs were divided into pointers, flushing dogs, and retrievers, and each were expected, against their natural instincts, to do one thing and one thing only: the pointers to point, the flushing dogs to flush, and the retrievers to retrieve. Their ability to stick to their assigned role and thus to be what we want them to be, *and nothing else*, is the measure of their goodness. We build machines that way, each for a particular purpose, and we measure their goodness in the same way, too: the good machine is the one that does reliably what it is supposed to do and nothing else. The role assigned to the animal is thus a role that could in principle be filled just as well by a machine. We only use animals because we haven't quite figured out yet how to build machines that can do what those animals do. Sometimes, it is easier to transform what already exists rather than to create something from scratch. So we make do with what we have got until something better comes along, something even more perfectly tuned to our ends, a cyborgised robodog perhaps.

Yet animal breeding practices have not always aimed at increasing the usefulness of animals for particular purposes. From the second half of the nineteenth century onwards, dogs and farm animals were increasingly bred for looks rather than use (Derry 48–66). Even today, strongly encouraged by national kennel clubs, dog breeders are still obsessed with arbitrarily contrived rules and measurements that are brought into action to determine the difference between a good (valuable and desirable) and a bad representative of its kind (Brandow 2015). The slightest deviation from the prescribed ideal proportions—which, in most cases, are completely unrelated to what an animal's health or even a sound body structure would require—makes the animal all but worthless. Such Vitruvian animals are only accidentally living and sentient beings. They are treated and regarded as carefully crafted artefacts, attesting to the power, imagination, and taste of their breeders and owners. Accordingly, dog breeding is often understood by its practitioners as an art rather than an applied science (Derry 2003, 15). Book titles such as *The New Art of Breeding Better Dogs* (Onstott 1975) are common. A reference guide on Great Danes informs us that the "breeding of dogs is truly an art. The artist's medium is living flesh" (Swedlow 1999). As an art form, dog breeding can, and indeed must, ignore any considerations of usefulness. Some of this fascination with the sheer ability to transform the phantasms of one's imagination into reality can also be found among transhumanists, for instance, when James Hughes (2004, 92) tells us: "When I was reading Harry Potter to my kids they asked if there really were unicorns, elves and centaurs. I told them no…not yet. But in the coming decades and centuries we will be able to create all the creatures that populate our mythologies." The Baconian utopia where the bounds of the human empire will finally be enlarged to such an extent that we can bring about "all things possible" (Bacon 1924) is here identified with the realisation of a child's fantasy world. It is a little boy's dream (or nightmare) come true. Of course, Hughes imagines those future fantasy creatures to have "human-level intelligence", which immediately raises the question whether we will then also have to assign citizen rights to them (yes, thinks Hughes). But just as the wolf has all but disappeared from the pure-bred pooch, the animal has disappeared from Hughes' vision, to be replaced by the intelligent, mentally human-like creatures that populate our fantasies.

At any rate, the enhanced animal is typically considered improved when it has become more like a human-made artefact, either by becoming more machine-like or more like a work of art. In doing so, the animal's well-being is largely ignored, or taken into account only to the extent that it is relevant for its functioning, which in turn is required for its optimal capitalisation. Animal bodies are simply, as Richard Twine puts it, "aligned to capital" (Twine 2013, 509). Even when animal welfare appears to be taken seriously, as something that counts in its own right and that can or should not be ignored, the idea of the better animal as a well-functioning machine often determines in what way the problem is sought to be solved. Thus, the American philosopher and animal rights campaigner Bernard Rollin (1995, 169–176) has suggested that, in order to increase animal welfare and to alleviate what he calls the "plight of the creature", we genetically modify farm and lab animals in such a way that they no longer suffer from the terrible living conditions that we force upon them. If there is no realistic hope to change those conditions (which there is not), then we should, if we can, just change the animal so that, for instance, chickens kept in battery cages are no longer miserable, but instead happily submit to the human goal of "efficient, high-yield egg production" (Rollin 1995, 172). If we did the same with humans, that would, according to Rollin, of course be wrong, but it is not wrong to do the same with animals because values like autonomy supposedly have no relevance here. Autonomy matters to us, not to them. In some cases, we even need to consider surgically rendering the animal decerebrate, "to obliterate all subjective experience, to totally eliminate consciousness", resulting in an animal that is "mentally dead but physically alive" (Rollin 1995, 205). The ethics of this *dumbing-down* approach has subsequently been widely discussed under the label of animal *dis*enhancement by various authors, including myself, but I am not going to revisit this debate here.[6] What is relevant in the context of this chapter's focus is that even when the professed rationale for a proposed modification is the interest or well-being *of the animal* (rather than human interests and well-being), the result is still the disappearance of the animal. It is assumed that what is best *for* the animal is that there is no animal.

It is noteworthy that Descartes, who, notoriously, denied that animals had a soul and that they actually felt pain or for that matter anything else even when it looked that way, based that denial, among other things also (and perhaps even primarily), on an *ethical* consideration, and many Cartesians followed him in this respect. If animals really felt pain, they argued, we would do them a terrible injustice by treating them the way we do because they clearly do not deserve any of it. Our behaviour towards them would therefore not be justifiable. Yet since God would not allow such an injustice to occur, we must, on ethical and theological grounds, believe that they do not really suffer when we torture and kill them (Rosenfield 1968, ch. 2). It is this kind of argument that Chesterton (1908, 24–5) had in mind when he quipped: "If it be true (as it certainly is) that a man can feel exquisite happiness in skinning a cat, then the religious philosopher can only draw one of two deductions. He must either deny the existence of God, as all atheists do; or he must deny the present union between God and man, as all Christians do. The new theologians seem to think it a highly rationalistic solution to deny the cat." For us, the Cartesian argument is no longer convincing, and today, we can no longer pretend that animals do not feel any pain, that there is, as it were, nobody inside them. We *know* that they are not machines (and that there indeed *is* a cat). But that does not mean we cannot *turn* them into machines. If it is not (yet) true that they are machines, then we can perhaps *make* it true by creating animals, or living things, that really *are* nothing but machines in the sense that Descartes imagined they were: functioning, but unfeeling. This is what Rollin's dumbing-down approach ultimately seeks to accomplish: the creation of a fully compliant beast-machine. Cartesian dualism still reigns supreme in this approach: the animal's consciousness, its ability to feel and its being a subject-of-a-life (Tom Regan), is understood as an expendable add-on that can, without loss, be eliminated from the body-machine, which is all we really need the animal to be. Alternatively, we may soon also be able to eliminate the animal (and thus spare them suffering) altogether by developing new ways of producing what we currently still need animals to produce,[7] for instance, by using tissue engineering to produce in-vitro meat. After all, the first lab-grown burger has already been cooked and eaten. If this became possible on a wide scale, then certain species of non-human animals might not only disappear in mind, but also in body.

The transhumanist proposal to "uplift" non-human animals follows the same trajectory. What it has in common with those other proposed ways to enhance or (disenhance) animals is the determination not to let the animal be what nature has made it. In one way or another, the unenhanced animal or the animal *qua* animal is always a nuisance. Thus, David Pearce (1995, sect. 1.10), in his eagerness to free the world and all sentient beings in it of all suffering, outlines his plan to turn all carnivorous animals into herbivores or, if that is not possible, to get rid of them altogether. It is the transhumanist version of the biblical prediction (if taken literally) of a coming golden age, when "the wolf and the lamb shall graze together; the lion shall eat straw like the ox, and dust shall be the serpent's food" (Isaiah 65:25). Except that it is less forgiving and more inclusive. Cats and other carnivores, declares Pearce, are in fact nothing but the animal equivalent of psychopaths (hence insane, once again). They are "pre-programmed killing machines" (which, apparently, is the wrong kind of machine), which we should not allow to continue to exist. In fact, it is our moral *duty* to make sure that they do not exist. Any desire to preserve them is nothing but a "misguided romanticisation". "In future," he says, "the life-forms which exist on this planet will be there purely because we allow them to be so, or choose to create them." Pearce realises that all this talk about *allowing* and *not allowing* living things to exist "smacks of hubris" but he is fine with that because he thinks it is both *true* and *right* that this is going to happen. His fellow transhumanist and fellow animal rights advocate George Dvorsky shares Pearce's unabashed "technovisionary paternalism" (Ferrari 2015), the conviction that we know best what is good for non-human animals, and indeed what is good and desirable in general and for everyone. Besides, we have the power, and with power comes responsibility, which we shouldn't shy away from. Hence, the uplift imperative. Dvorsky defines animal uplifting—a term borrowed from David Brin's 1980's *Uplift* novels—as "the theoretical prospect of endowing nonhumans with greater capacities, including and especially increased intelligence" and claims that we "are morally obligated to biologically enhance nonhuman animals and integrate them into human and posthuman society" (Dvorsky 2008, 130, 129). The assumption behind the postulated "ethical imperative to uplift" is that a non-human animal's life generally resembles more a Hobbesian nightmare than a Rousseauian Garden of Eden: it is "nasty, brutish and short". Also, they lack political participation and what comes with it, namely liberty and justice. By uplifting them to a human (or, if we also uplift ourselves, post-human) level of intelligence, we would empower non-human animals "to participate in the broader social community" (137) and to live "a more dignified and fulfilling life" than is currently, due to the limitations of their nature, available to them (132). Uplifting will allow both us and them to transcend those biological limitations. However, since those limitations are much more severe and inflexible for them than they are (normally) for us, so that they never really reach "minimally acceptable modes of functioning", non-human animals can be "construed as disabled humans" (138). Dvorsky thus adopts and reaffirms Hughes' disability narrative. He does *not* say (although it seems to follow logically) that disabled humans can then also be construed as animals. The whole point of the comparison is, after all, to associate animals with a state of disability, and not to associate human disability with an animal-like state. (And, as we will see and explore in the next chapter of this book, disability is a condition that, in the transhumanist narrative, we *all* share, whether human or non-human, conventionally disabled or conventionally abled.) The term "disability" suggests not only an absence, but also the absence of something that *should* be there. Non-human animals lack something important that *we* have.

Dvorsky, however, denies that the uplift project is in any way anthropocentric. If anything, he says, it is "intellicentric and even quasi-perfectionist" (2008, 138). The human is, after all, not the final stage of the uplifting process, but at best a transitional stage. Uplifting is not about making animals more like us, but about not leaving them behind when we move on to a higher level of consciousness and existence. Ultimately, we will all, humans and animals alike, be "post-biological" rather than merely post-human, post-ape, or post-elephant. What we once were will then be irrelevant. This post-biological state is the transhumanist heaven: a state where all traces of our former animal nature have been erased, and hence, a state of complete autonomy. It is then that we can finally be what we were always meant to be. It is not going to be a human life, if by human we mean a being with a certain, namely human, biology, a member of the (animal) species *Homo sapiens sapiens*, but it is an actualisation of what we (or at any rate transhumanists) take ourselves to *truly* be, the ghost in the machine, the *sapiens* in the *homo*. It is this *ideal* humanity, a humanity that is liberated from their mortal shell (the shell that *makes* us mortal), that we generously allow the uplifted animal to partake in.[8] The expected state of complete liberation from our biological constraints is conceived as *intrinsically* desirable, which implies that it must be good for non-human animals, too. Yet since the animal will no longer *be* an animal when it is in that state, it follows that what is seen as good for the animal is not to be an animal at all (and for the human *qua* animal not to be a human). To make sense of this, we must of course assume that we can separate the animal from what it is as an individual, or in other words, that it being a particular kind of animal is accidental to its existence, so it can still be itself even when it is no longer that animal, or not even an animal at all. If that is not plausible, then animal uplifting is not an *enhancement* of an existing animal *to* a post-animal, but a *replacement* of the animal *with* a post-animal. The animal goes extinct to make room for the post-animal—just as the human goes extinct to make room for the post-human. Nick Bostrom's quirky tale about a Golden Retriever named Albert who first gets uplifted and mind-uploaded and is then interviewed on the Larry King show (Bostrom 2004) is a case in point. As a post-biological being, the post-dog Albert is nothing like the animal that he used to be. He may still be Albert, but he is certainly no longer a dog. Nicholas Agar (2014) has argued that post-humans will be so radically different from us that they will soon forget what it was like to be us. They will also have very different interests and values. For this reason, we would have nothing to gain by becoming post-human, even if being post-human were intrinsically and instrumentally good. Radical enhancement may be good for the post-human, but it is not good for us, simply because they will no longer be us. As long as we only enhance ourselves modestly, we will remain the same person, but once we have undergone radical enhancement, we will have become something else entirely. For post-dogs and other non-human post-animals, the gap between what they are and what they used to be will even be larger than the gap between the post-human and the human, which will only exacerbate the identity problem (although if the post-human/post-animal is advanced and remote enough, then the difference between human and animal may become negligible). In this respect, animal uplifting is comparable to the, if you will, downdragging that happens to Gregor Samsa, the protagonist of Franz Kafka's short story "Metamorphosis" (1915), who wakes up one morning to find himself transformed into a giant cockroach-like creature. Of course, that cockroach is rather unlike the radically enhanced post-human as we tend to imagine them. But it is definitely the product of a sudden radical transformation of a human being into something that is (in some, though not all respects) very different from what it used to be. The transformation does not only affect Gregor's body, but also his interests, appetites, sensibilities, and needs. At one point in the story, Gregor realises that he finds it increasingly difficult to remember his life prior to his transformation and how it was to be human, and to still care for what he cared for while he was still human. And that is perfectly plausible: the world of a giant cockroach, even one that used to be human, is so different from that of a human that we can understand how everything human must fade away very quickly in such a creature's mind. A cockroach's body demands (and creates) a cockroach's mind. So if the transformation of a human into a radically enhanced human or post-human, or of an animal into a post-animal is anything like Gregor Samsa's transformation into a giant insect, then we should indeed expect, as Agar argues, that they will very soon no longer remember much, or anything at all, of their previous existence. This is of course not necessarily a good reason not to pursue such a transformation. I suspect that the reason why we would not want to find ourselves in the same predicament as Gregor Samsa is not that he forgets his humanity, but rather that he has turned into an animal that we tend to find disgusting and that is not known for its superior intellect (although we may have to revise this judgement in the light of our failure to fool the roboroach for more than just a few minutes). Perhaps we would not mind much losing connection to our previous self if what we gain in exchange for it is a richer and (intrinsically and extrinsically) more rewarding life.

This is exactly what transhumanists believe that uplifting would do: give non-human animals a better, nobler life. While we may not have it yet, we at least know that there is such a life and that it is worth striving for. Animals don't even know that there is a better life and certainly cannot appreciate its value. The very word "uplifting" suggests a hierarchy, a difference between lower and higher states of existence. We can only *uplift* what is on a lower level, and *we* can only do the uplifting if we are on a higher level already (which does not preclude the possibility of even-higher-than-human levels). Dvorsky (2012) approvingly cites David Brin, the author of the *Uplift* saga and, like Dvorsky and Hughes, fellow at the Institute for Ethics and Emerging Technology, who, in an interview, accuses evolution of being "stingy" for not letting non-human animals crash through the "firm glass ceiling" of limited abilities under which they are stuck. (We, on the other hand, have, somehow, crashed through, although we may have a, somewhat less firm, glass ceiling of our own.) It would, he says, be selfish of us to let them stay there and to keep the benefits of enhancement technologies to ourselves. "Imagine dolphin philosophers, bonobo therapists, raven playwrights and poets," he says, "How lonely, if we turn away without trying." That we might be *lonely* without uplifted animals is a curious and telling worry. It assumes that we cannot communicate with non-human animals, that they live in one world and we in quite another. That we cannot communicate with them in our language (i.e. a language that *we* can understand) is clearly perceived as frustrating. It is yet another limitation imposed on us. There are worlds of experiences out there that we cannot grasp, that are closed to us. We have no idea at all what it is like to be a bat. Or for that matter a dog.[9] Uplifting will change that: it will finally allow us to know what it *is* like. Except that the uplifted bat is no longer a bat, the dolphin turned philosopher no longer a dolphin, and the raven turned poet no longer a raven. Once they have been enabled to communicate with us in our language, they are no longer the kind of otherworldly being that we wanted to communicate with in the first place. "If a lion could speak," Wittgenstein (1953, 223) remarked, "we could not understand him." We will, however, understand the post-lion, precisely because he will no longer be a lion, which is just as well. With only post-animals around (since we will not *allow* any unenhanced animals to exist), we will no longer be constantly reminded of our limitations because there is nobody left that we cannot communicate with, nobody who defies our understanding and is beyond our reach, beyond our control. The animal is that which cannot be controlled (and the animal *in* us, the animal that we are, is everything in us that we cannot control). Giving non-human animals human-like mental abilities is a way to make them less alien and more compliant. The autonomy that is bestowed on them is a form of appropriation. Uplifting is less about giving non-human animals a mental form that finally makes them deserving of equal moral recognition (as Hughes seems to think), but about giving them what they need to recognise us: as their creators, saviours, and ultimately, superiors. In *Sundiver*, the first instalment of Brin's *Uplift* trilogy (2012), first published in 1980, an argument ensues between a human and an uplifted chimp technician called Jeffrey. When Jeffrey gets mad and physically attacks the human in an ape-like fashion, another human, the novel's main protagonist, a man called Jacob, intervenes: "Jacob took the chimp's face in his hands. Jeffrey snarled at him. 'Chimpanzee-Jeffrey, listen to me! I am Jacob Demwa. I am a human being. I am a supervisor with Project Uplift. I tell you now that you are behaving in an unseemly manner…you are acting like an animal!' Jeffrey's head jerked back as if slapped." When a chastened Jeffrey apologises to his human opponent, Jacob praises him: "'That's fine,' Jacob said. 'It takes a real man to apologize'" (Brin 2012, 67).

The ex-animal apologises for behaving like an animal. The uplifting process was meant to civilise and discipline it, and when it falls back to its animal ways, it needs to be disciplined again by being reminded of its status, its precarious and paradoxical position as the animal-it-was-but-no-longer-is. It is thus not surprising that uplifting can, as Dvorsky (2012) acknowledges, "be construed as being imperialistic and over-domineering – an unfair and unwarranted imposition of 'humanness' onto the animal kingdom". Yet Dvorsky's concession that "there's something to be said for living in an innocent state of mind – even if it is in the jungle" rings false. The phrasing betrays the same condescending attitude towards real pre-enhancement animals that informs the whole uplift project. The animal's "innocence" is just a euphemism for an absence of (human-like) knowledge and understanding, which a transhumanist cannot but find deplorable. For the transhumanist, innocence means ignorance, and ignorance is bad. That kind of innocence is quite compatible with Pearce's assessment of carnivorous animals as psychopaths. And the "jungle" indicates a nature that is red in tooth and claw, untamed, uncivilised, unpredictable. This jungle is clearly not a paradise. It is a place that we cannot imagine anyone would like to stay in if they had the choice to leave it. I'm an animal…get me out of here.

So that is what transhumanists urge us to do: get the beast out of the jungle, make it presentable. I find myself reminded of yet another of Kafka's stories, "Report to an Academy", published almost exactly a century ago (1917). In that story, a former ape reflects on his transformation from ape to human-like post-ape and explains why this transformation has occurred. Red Peter—as human society has dubbed him—lived his life as a free ape until he was shot and captured by hunters, who teach him how to drink alcohol and how to spit. He finds himself crammed into a small cage, made fun of, and occasionally tortured. He knows that even if he manages to escape, it would do him no good because he would only be captured again. So he reasons that if this is the place that an ape has to live in, then there is only a way out for him if he stops being one, and becomes human. So he observes and imitates, learns to speak like a human, and to act like one, until he is finally human enough to be allowed to live a human-like life in a human world. By adopting human ways, he has managed to survive and to get out of the cage. He has not, however, acquired freedom. Freedom, he says, is something that he perhaps had (he cannot quite remember) when he used to be an ape and what some humans may yearn for. That freedom he has not regained by submitting himself to "the yoke" of human civilisation.

This suggests that there are two different kinds of freedom. One is the self-regulating autonomy that characterises modern human life and that transhumanists seek to expand and extend to non-human animals, ultimately aiming at liberation from all biological constraints. The other is the freedom of the jungle that any wild animal still has and that we humans have mostly lost. This is the freedom to live one's life as the kind of creature that one is, without the pressure or need to change and become something else. Like Kafka's Red Peter, animals may only want to choose the former if they have no other way out: if ceasing to be what they are is the only chance they have to be left in peace and not to be subjected to our human needs and wants.

Notes

1. Although Haraway has later expressed reservations about being called a posthumanist: "I never wanted to be posthuman, or posthumanist, any more than I wanted to be postfeminist" (Haraway 2008, 17).
2. Jethro Knights, the hero of Zoltan Istvan's transhumanist novel *The Transhumanist Wager*, chides an assembly of academics and politicians who are less than thrilled by the transhumanist agenda by angrily exclaiming: "Do we really want to remain animals for the rest of our days when we can be so much more?" (Istvan 2013, 28).

3. Donna Haraway (2008, 79) remarks, with reference to Derrida, that "patricide and fratricide are the only real murders in the logic of humanism". Matricide, on the other hand, is permissible.
4. "We must stand guard against our genes, lest they chain us to remaining as animals forever", declares Jethro Knights in *The Transhumanist Wager* (Istvan 2013, 278): "Biology is for beasts, not future transhumanists."
5. Just as the transhuman is a human who is already in the process of becoming something other or more than human, that is, post-human, the *transanimal* is an entity already in the process of becoming more than animal, or postanimal. The extension of these two terms is the same. The transhuman of transhumanism is identical with the transanimal. Accordingly, its post-human is identical with the postanimal.
6. See, for instance, Hauskeller (2007, 117–128), Thompson (2008), Palmer (2011), Henschke (2012), and Ferrari (2012).
7. I don't mean to suggest here that we need meat to live or live well. Vegetarianism and even veganism is certainly a viable option. All I am saying is that *if* we want to eat meat, we currently need animals to provide it. If we could grow meat in the lab, this would change.
8. Although, as Celia Deane-Drummond (2011, 118–9) has pointed out, it is highly likely that animals will *first* have to serve as test subjects for the enhancement of humanity.
9. There is a new device that promises to change that: *No More Woof*, produced by the Norwegian start-up Indiegogo (https://www.indiegogo.com/projects/no-more-woof#/). The device is attached to the dog's head and uses micro computing and electroencephalogram (EEG) to analyse the dog's thought patterns, which are then translated into English, allowing the dog to inform us that it is hungry, or tired, or wants to go for a walk (just in case their human owners are too stupid to understand the dog's body language). This is thought to be only the beginning.

Bibliography

Agar, Nicholas (2014) *Truly Human Enhancement: A Philosophical Defense of Limits*, Cambridge, MA: MIT Press.

Bacon, Francis (1924) *New Atlantis*, Oxford: Clarendon Press.

Bostrom, Nick (2004) "Golden", www.nickbostrom.com.

Brandow, Michael (2015) *A Matter of Breeding*, Boston: Beacon Press.

Brin, David (2012) *Uplift. The Complete Original Trilogy*, London: Orbit.

Chesterton, Gilbert Keith (1908) *Orthodoxy*, London: Bodley Head.

Deane-Drummond, Celia (2011) "Taking Leave of the Animal? The Theological and Ethical Implications of Transhuman Projects", in *Transhumanism and Transcendence*, ed. Ronald Cole-Turner, Washington, DC: Georgetown University Press, 115–130.

Derry, Margaret E. (2003) *Bred for Perfection. Shorthorn Cattle, Collies and Arabian Horses since 1800*, Baltimore/London: John Hopkins University Press.

Dvorsky, George (2008) "All Together Now: Developmental and ethical considerations for biologically uplifting nonhuman animals", *Journal of Evolution and Technology* 18/1: 129–140.

——— (2012) "Should we upgrade the intelligence of animals?", *io9 blog* 9.7.2012, http://io9.gizmodo.com/5943832/should-we-upgrade-the-intelligence-of-animals.

Ferrari, Arianna (2012) "Animal Disenhancement for Animal Welfare: The Apparent Philosophical Conundrum and the Real Exploitation of Animals", *Nanoethics* 6: 65–76.

——— (2015) "Animal Enhancement: Technovisionary Paternalism and the Colonisation of Nature", in *Inquiring into Animal Enhancement: Model our Countermodel of Human Enhancement?*, eds. Simone Bateman et al, Basingstoke: Palgrave Macmillan, 13–33.

Haraway, Donna (1985) "A Manifesto for Cyborgs. Science, Technology, and Socialist Feminism in the 1980s", in: *Feminism/Postmodernism*, ed. L. Nicholson: New York and London: Routledge 1990, 190–233.

——— (2008) *When Species Meet*, Minneapolis/London: University of Minnesota Press.

Hauskeller, Michael (2007) "A Legacy of Love? Reflections on the New "Humanitarian" Eugenics," in *Utopie heute 1*, ed B. Sitter-Liver, Fribourg: Academic Press/Stuttgart: Kohlhammer, 405–419.

——— (2013) *Better Humans? Understanding the Enhancement Project*, London: Routledge.

Hayles, N. Katharine (1999) *How We Became Posthuman*, Chicago and London: The University of Chicago Press.

Henschke, Adam (2012) "Making Sense of Animal Disenhancement", *Nanoethics* 6: 55–64.

Hughes, James (2004) *Citizen Cyborg. Why Democratic Societies Must Respond to the Redesigned Human of the Future*, Cambridge, MA: Westview Press.

——— (2013) "Transhumanism and Personal Identity", in *The Transhumanist Reader*, eds. Max More and Natasha Vita-More, Chichester: Wiley Blackwell, 227–233.

Humanity Plus (1998) *Transhumanist Declaration*, http://humanityplus.org/philosophy/transhumanist-declaration/

Istvan, Zoltan (2013) *The Transhumanist Wager*, Futurity Imagine Media.

Kafka, Franz (1915) *Metamorphosis*, https://www.gutenberg.org/files/5200/5200-h/5200-h.htm

——— (1917) *A Report for an Academy*, http://records.viu.ca/~johnstoi/kafka/reportforacademy.htm

Kurzweil, Ray (2005) *The Singularity is Near*, New York: Viking Adult.

Moravec, Hans (1988) *Mind Children. The Future of Robot and Human Intelligence*, Cambridge, MA: Harvard University Press.

More, Max (1995) *The Diachronic Self: Identity, Continuity, Transformation*, Ann Arbor, MI: A Bell and Howell Company.

——— (2013a) "The Philosophy of Transhumanism", in: *The Transhumanist Reader*, eds. Max More and Natasha Vita-More, Chichester: Wiley-Blackwell, 3–17.

——— (2013b) "A Letter to Mother Nature", in: *The Transhumanist Reader*, eds. Max More and Natasha Vita-More, Chichester: Wiley-Blackwell, 449–450.

More, Max, and Natasha Vita-More (eds) (2013), *The Transhumanist Reader*, Chichester: Wiley Blackwell

Onstott, Kyle (1975) *The New Art of Breeding Better Dogs*, New York: Howell Book House.

Palmer, Clare (2011) "Animal Disenhancement and the Non-Identity Problem: A Response to Thompson, *Nanoethics* 5: 43–48.

Pearce, David (1995) *The Hedonistic Imperative*, www.hedweb.com.

——— (2007) "The Abolitionist Project", www.abolitionist.com.

Rollin, Bernard (1995) *The Frankenstein Syndrome. Ethical and Social Issues in the Genetic Engineering of Animals*, Cambridge: Cambridge University Press.

Rosenfield, Leonora Cohen (1968) *From Beast-Machine to Man–machine. Animal Soul in French Letters from Descartes to La Mettrie*, New York: Octagon Books.

Sandel, Michael (2007) *The Case Against Perfection. Ethics in the Age of Genetic Engineering*, Cambridge, MA: Harvard University Press.

Sharon, Tamar (2014) *Human Nature in an Age of Biotechnology*, New York: Springer.

Swedlow, Jill (1999) *Great Dane. Model of Nobility*, New York: Howell Book House.
Thompson, Paul B. (2008) "The Opposite of Human Enhancement: Nanotechnology and the Blind Chicken Problem", *Nanoethics* 2: 305–316.
Twine, Richard (2013) "Animals on Drugs: Understanding the Role of Pharmaceutical Companies in the Animal-Industrial Complex", *Journal of Bioethical Inquiry* 10/4: 505–514.
Wittgenstein, Ludwig (1953) *Philosophical Investigations*, Oxford: Basil Blackwell.
Wolfe, Cary (2010) *What Is Posthumanism?*, Minneapolis: University of Minnesota Press.

7

The Disease of Being Human

In the transhumanist mythology, the animal that needs fixing most of all is the animal that we are ourselves. Being an animal means to be disabled, to not be able to do what we would need to do to live a truly good life. This is a message that these days we have come to be quite receptive to. It is now getting increasingly difficult not to be permanently confronted with visions of a technologically enhanced humanity. Human enhancement is the latest fashion. It is exciting, enticing, cool, and sexy. Philosophers fantasise about the wonderful lives we are all going to enjoy once we have shed our mortal shell and become post-human (which, it is believed, will be very soon), and the media are eager to spread the good tidings and do their best to whet our appetite for our own terminal transformation into something very different. If transhumanism is a philosophy that endorses and promotes radical human enhancement, then it seems that we are all transhumanists now. This means more than just being open to change. It involves a determination to propel us forward into the future, driven by the deep conviction that the present condition of humanity is utterly deplorable and a diseased state. If the human condition is the primary

M. Hauskeller, *Mythologies of Transhumanism*,
DOI 10.1007/978-3-319-39741-2_7

disease, then radical human enhancement is the cure. This implies that radical human enhancement is more than just an option: it is a moral obligation. To be a transhumanist means to be a healer of humanity.

Two decades ago, in the mid-1990s, when the ethical debate about human enhancement really took off, one of the most contentious issues was whether one could meaningfully distinguish between therapy and enhancement and whether that distinction was ethically relevant.[1] Although the distinction and its ethical relevance was already contested by proponents of the human enhancement project, those who had their doubts about the project and the desirability of its projected outcomes had not given in yet and were still putting up a fight—and the public was largely on their side. The issue was important because it was generally assumed that therapy was a good thing, something worthy of being supported and endorsed, clearly permissible, and most likely even something that we *owe* to people, something that everybody has a right to be provided with. Hence, if no clear distinction between therapy and enhancement could be made, then there was no reason to reject enhancement procedures as unethical or undesirable. On the contrary, we might even have good reason to demand general support for it and a commitment by the state to actively pursue the development of enhancement technologies, with the goal of eventually making them available to all citizens. On the same grounds, those who believed that human enhancement was *not* desirable and certainly not something we have a moral duty to bring about had every reason to hold on to the distinction and to defend its validity.

Today, the alleged distinction between therapy and enhancement is no longer an issue. The debate has moved on. So has the public perception, which no longer seems to recognise or appreciate the difference. When I google the terms "enhancement and therapy" today, the vast majority of entries that I find are not concerned with the possible distinction between the two concepts, but instead seem to assume that enhancement is *itself* a form of therapy. Thus, there are plenty of entries offering and discussing *enhancement therapies*, such as "motivational enhancement therapies", "cognitive enhancement therapies", "relationship enhancement therapies", "spiritual enhancement therapies", and so forth, indicating that, as far as the public understanding of these terms is concerned, the erstwhile fight to uphold the difference has clearly been lost. We are now perfectly happy to incorporate the concept of enhancement in our concept of therapy.

There are various argumentative strategies that one can use to justify this move. We can try to *blur* the difference by, for instance, citing real-life cases that we find difficult to subsume under either of the two categories. Is an intervention that allows a 70-year old woman to conceive and give birth to a child enhancement or therapy? Or one that allows a 70-year-old man to have the love life of a 20-year old? Is it *therapy* to treat a boy who suffers from growth hormone deficiency, but *enhancement* to treat one who just happens to be shorter than the average boy of his age? Is vaccination enhancement or therapy? It does not seem to make much sense to insist on the difference between enhancement and therapy in such cases. Another strategy consists in acknowledging the difference (in most, if not all, cases) but denying its ethical relevance: we should not be interested in whether some intervention is classed as enhancement or as therapy (though it may well be possible to do so), but rather focus on whether it is *beneficial* to us. If the whole point of therapy is to improve a patient's well-being and something that we would normally call enhancement does exactly the same, then it does not really matter what we call it, we should definitely support it.

A more radical and ingenious strategy to gain support for the enhancement project than that of blurring the difference between therapy and enhancement or denying its moral relevance is to maintain that enhancement actually *is* nothing but (an extension or a particular kind of) therapy. There are two main theoretical arguments to support this claim. I call them the *moral argument* and the *biological argument*. The moral argument is mainly associated with the British philosopher and bioethicist John Harris, and the biological argument with the American philosopher and bioethicist Allen Buchanan.

Already in 1993, long before he used it as the cornerstone of his unreserved, crypto-transhumanist endorsement of radical human enhancement in his 2007 book *Enhancing Evolution*, Harris presented the moral argument in a paper, published in the journal *Bioethics*, that was, judging by its title, meant to answer the question whether "gene therapy [was] a form of eugenics". Harris's answer was that it may well be, but that it did not matter. What matters is whether it is desirable, and on reflection, it should be clear that gene therapy (by which he meant genetic *enhancement*) is not only desirable, but indeed morally obligatory. We all have a duty to support the development and use of human enhancement technologies because there is no relevant moral distinction between *repairing a dysfunction* and *enhancing a function*, so if the former is a duty, then the latter is too. This conclusion is reached through a seemingly logical progression from certain assumptions that nobody is likely to deny. Surely, Harris argues, every parent has the right to wish for "a fine healthy child" and to do everything in their power to make sure that their child will indeed be fine and healthy. But equally, we would consider it *wrong* for a parent *not* to do everything in their power to secure that outcome and to prevent their child from being born disabled or in any other way harmed. It would be wrong for the simple reason that *not preventing* a disability or harm is tantamount to *causing* the condition. This claim is based on the so-called *moral symmetry principle*, which is widely accepted among bioethicists (at least those of a broadly utilitarian persuasion). The moral symmetry principle was proposed by Michael Tooley in 1972. It suggests that if the outcome is the same, then action and inaction are morally on a par, meaning that not preventing an evil is morally as bad as actively causing it. For instance, you are just as responsible for the death of a child that you do not prevent from drowning when you could easily have saved them (without endangering yourself) as you would be if you had drowned that child yourself. Yet in order to reach the desired conclusion that standing in the way of human enhancement is morally wrong (and indeed evil), something more is needed, namely a new definition of what it means to be disabled, and this is exactly what Harris provides. Disability, he claims, is "a condition we have a strong rational preference not to be in" and one that is "in some sense a harmed condition" (Harris 1993, 166). And what is a "harmed condition"? It is one that (once again) someone has a strong rational preference not to be in and which might be described as harmful, "not relative to normal species functioning but *relative to possible alternatives*" (Harris 2007, 92, my emphasis). In other words, disability (i.e. a condition that it would be wrong not to try to rectify) is redefined as harmed condition and, more importantly, harm is redefined as unnecessary (given that there are alternatives) disadvantage.

It is clear that these are both persuasive definitions in Stevenson's sense, intended to preserve the emotional meaning of the terms (disability and harm are both *bad*), while changing the conceptual content, thus redirecting the way we look at what was previously understood as normal and hence not in need of a cure. The new definitions have two advantages. For one thing, it is suggested that you can be harmed without even being aware of it. No subjective suffering is required for you to be in a harmed condition. And for another, we are asked to accept that what constitutes harm (and its opposite, health or well-being) is entirely contextual and comparative, depending on what is medically and technologically possible at a given time. It follows that the mere possibility of certain enhancements (the fact that they are already available to us or even that they *could* conceivably *become* available to us if we invested enough money and research) is sufficient to render the unenhanced state that we are currently in, that is, the familiar human condition, a state of unnecessary disadvantage and thus of harm. Take, for instance, the possibility of radically extending human life span (which some, like the British gerontologist Aubrey de Grey, believe is imminent, claiming that the first person to live for a thousand years is likely to be already alive today). According to Harris, if "the gene therapy could enhance prospects for healthy longevity then just as today, someone who had a life expectancy of fifty years rather than one of seventy would be regarded as at a substantial disadvantage, so having one of only seventy when others were able to enjoy ninety or so would be analogously disadvantageous" (Harris 1993, 168).

For Harris, "'death postponing' is after all just 'life saving' redescribed" (1993, 168). In a world in which technology promises to makes us even more abled than we normally are, the normally abled become the disabled. In the face of what we could be, the difference between those of us who possess the normal human abilities and those who lack some of them (and who would, for this reason, normally be regarded as "disabled") becomes negligible. The mere possibility of super-ability makes us *all* disabled. And since this is clearly bad, so the argument goes, we are morally obligated to do something about it. We are morally obligated to enhance ourselves and our children.

In contrast to the moral argument put forward by John Harris, the biological argument proposed by Allen Buchanan does not appeal to a particular understanding of moral obligation. It does, however, also rely on a persuasive redefinition, or at least reinterpretation, of what it means to be in a harmed (disabled or diseased) state (and thus in need of a cure). Buchanan attacks what he takes to be the conventional view of nature (or evolution) as some kind of master engineer and, accordingly, of its products (living beings, which includes us humans) as finely tuned masterworks of creation, in which everything has a purpose and everything fits together perfectly with everything else (Buchanan 2011, 155–161). According to Buchanan, nothing could be further from the truth. Far from being a master engineer, nature is more like a blundering, blind fool or, at best, a mere tinker, a rather incompetent amateur who tries his best with the materials that he happens to find in his garden shed to create something that does not fall apart right away but which is certainly not good enough to last very long. Living organisms may indeed be "finely tuned", but they are so in the same way a house of cards may be said to be finely tuned in the sense of being perfectly balanced. This only means that it may be brought down by the slightest disturbance. The truth is that we are very poorly designed, which means that human enhancement, that is an improvement of our very nature, our design as a living organism, is more than just desirable. It is urgently needed to safeguard nothing less than our own survival as a species. "If the human organism is so poorly designed as to be exceedingly fragile, then we may need to improve it if we are to survive" (Buchanan 2011, 158). This is a reinterpretation of what it means to be in a harmed condition because it assumes that harm (i.e. an intolerable and, in the long run, fatal weakness) is an in-built characteristic of our species and indeed the very essence of our human condition. If our very humanity is the primary disease, the disease at the root of all diseases, then harm is no longer an exception, a deviation from the normal, healthy, and unharmed state of being. It is identical with what we are, which means that we are all, simply by virtue of being human, in need of a cure.[2] Human enhancement is then nothing but the ultimate therapy. If you believe that, then it is difficult to avoid the kind of practical conclusion that Max More (1996) proclaimed at the end of his seminal paper "Transhumanism: Towards a Futurist Philosophy": "Let us blast out of our old forms, our ignorance, our weakness, and our mortality. The future is ours."

Applications of this view, which has been argued for, although in different ways and on different grounds, by both Harris and Buchanan, and which we may call the *enhancement–therapy identity thesis*, can be found abundantly both in the current academic literature on human enhancement and in the popular culture which receives and reflects it. This increasing tendency to unquestioningly accept the enhancement–therapy identity thesis is part of a process that I like to think of as a transhumanisation of our culture. In the following, I briefly provide some examples of discourse relating to the four main (i.e. most widely discussed) areas of human enhancement: emotional enhancement, cognitive enhancement, moral enhancement, and life extension. In each of these cases, we can identify a *diagnosis* relating to the supposedly intrinsically pathological human condition and a proposed *cure* that consists in the successful execution of some form of capacity enhancement.

Emotional Enhancement

Diagnosis: We are all enslaved by our emotions.

Cure: Using human enhancement technologies to gain control over our emotions.

Emotional enhancement is not only about making people happier (and thus increasing their subjective well-being), but also, and perhaps even more so, about being able to adjust our moods and emotions to what we think is *required* in a particular situation and to how we think we *should* feel in line with our own best (second-order) *interests* and rational *goals*. (It may of course also be about how we think *others* should feel to best serve *our* interests and goals, rather than theirs.) If, for instance, it is required that we feel sad, rather than happy, then we should be able to *be* sad. If we feel we should be angry, then we should be able to *be* angry. It is a matter of control, of being the master in our own house, of our own body and mind. That is what we hope to gain through mood enhancement. The current state of affairs, in which we are *not* in control, or only to a very limited degree—we can perhaps choose to do things that are likely to make us happy or sad, but in most cases, whether we are happy or sad will depend on what *happens*, be it in the world out there or in our own body, rather than on what we *want*—, is regarded not only as unsatisfactory *to us*, given our desires and interests, but also as deeply wrong in itself, as *intrinsically* or *objectively* defective.

In accordance with this assessment, Brian Earp (who is *not* a transhumanist), with some of his colleagues from the Oxford Uehiro Centre for Practical Ethics, has recently suggested that we should see human love and pair-bonding as an *addiction*, mostly based on the strong (physiological and psychological) similarities that could be found between drug or alcohol addiction and "love- and sex-based interpersonal attachments" (Earp et al. 2016). While in previous publications, Earp and colleagues only argued that *some* forms and instances of love may be considered bad for the people involved (i.e. detrimental to their well-being) and should therefore be treated with love enhancement technologies (should those become available) (e.g. Earp et al. 2013), they are now, by emphasising the addictive nature of love, strongly suggesting that love *itself* is a disease, something that we need to be cured of. Even though they assure the reader that they are not proposing that we try to eradicate *all* love (but once again only those forms and instances that compromise people's well-being), they do nonetheless imply as much when they maintain that, by its very nature, love is usually bad because it generally involves "despair, desperate longing, and the extreme and sometimes damaging thoughts and behaviors that can follow from love's loss" (Earp et al. 2016). If all love is likely to compromise our well-being, then we may well decide that we would be better off without it. Thus, the diagnosis of a pathology is extended from certain kinds of love to love itself. The suggestion that we should differentiate between (a desirable) medicalisation and (an undesirable) pathologisation of love, which Earp and colleagues have made elsewhere (Earp et al. 2015), does little to change this picture. The goal is still to enable us to modulate the physiological and neurological processes that underlie human love and relationships and thus to learn how to control the way we emotionally relate to other people: whether we feel attracted or attached to them or not, when, and to what extent. The proposed "medicalisation of love" serves this purpose, while supposedly avoiding the implication that love, or certain instances of love, are *diseases*—which would then, in turn, create or greatly increase social pressures to rectify the situation. According to Earp and colleagues, the pathologisation of a condition would increase the danger of "oppressive normalization and top-down control", whereas mere medicalisation (even pharmaceuticalisation) would not (or need not?) have that effect because all that medicalising that condition means is that we would see it as one that, in a particular individual's case, merits medical treatment, which is to say that we see it as something that diminishes that individual's quality of life. "Treating" the condition would then not mean *curing a disease*, but simply *improving well-being* by means of medical technology, which we can do without having to identify the treated condition as a disease with objective clinical–pathologic indices. In other words, if you are in love and your being in love gives you trouble (or you are *not* in love and your *not* being in love gives you trouble), that is, if your love-related feelings (or the lack thereof) make your life less good than it would be if those feelings were different, then there would be nothing really wrong with you. Yet if you could change that situation through a particular medical intervention, then it would be absolutely fine for us to provide you with that opportunity, and for you to seize it.

However, the strict separation between treatment and pathology, which Earp and colleagues believe would "further diffuse the potential problem of the pathologization of everything", may well have the exact opposite effect. If well-being and the advancement of well-being is all that counts and all that medicine should concern itself with (thus eliminating the distinction between therapy and enhancement, or rather making any such distinction entirely moot), then we no longer have to bother trying to identify particular conditions as states of disease to justify their "treatment". Instead, *everything* we do and everything we are can now be regarded as fair game for medical interventions. As long as well-being, or our "quality of life", can be further improved—and it is hard to imagine any human condition where that is not the case—there is nothing about us that would *not* fall under the remit of medicine. From here, it seems a small step to declaring the whole human condition to be a disease that is in urgent need of a cure that only radical human enhancement can provide. The problem with a medical focus on well-being is that well-being is not clearly defined. We can never be sure that we are well enough. Our lives can always conceivably be better than they are. If we are happy, there is still a possibility that we can be even happier. If we love someone, it is still imaginable that we may love them even more, or more intensely, or less selfishly, or in some other way better. And even if we rather arbitrarily declared certain emotional states to be *good enough* (which would not go well with the inherent logic of the human enhancement project), it would be difficult to identify a state of love that does not at least have the potential to conflict with our well-being. Loving someone always holds a risk. It makes us more open to certain kinds of suffering. What Earp and colleagues thus propose is that we find a way to keep all the good stuff that comes with love while getting rid of all the bad stuff, and perhaps that is just not possible because it would change the very nature of love, part of which is that it is *not* under our control (or only to a small degree) and that it makes us more vulnerable by creating the possibility of devastating loss. Of course it may be argued that if *that* is what love is, then we would be much better off without it anyway. Clearly, although interpersonal relationships are indeed important for our well-being, so there are medical benefits to improving our relationships, for the purpose of enhancing well-being we could just as well try to find ways to make interpersonal relationships less important to us. A rational risk assessment may lead us to the conclusion that by far the best solution to the problem of love-related troubles is, once again, to get rid of love altogether. As we will see in Chap. 10, this is pretty much what transhumanists would have us do.

Cognitive Enhancement

Diagnosis: We are all stupid, suffering from various, hardwired cognitive deficiencies.

Cure: Using human enhancement technologies to increase our brain power.

Smart drugs such as modafinil have long been used by students to improve their performance in exams, but their use is now increasingly becoming a lifestyle choice. The general view seems to be that we could all do with a little more smartness, focused attention, alertness, and wakefulness to meet the rising demands of our busy lives. We are taught to feel deficient, mentally challenged, simply not up to the task. And it is not only our professional lives that are affected by this acquired lack of confidence in our own ability to cope, equipped as we are only with the limited intellectual powers that we possess by nature, but increasingly, *life itself*. Thus, Provigil (modafinil) is marketed with the slogan "Wake up to life", suggesting that we have been asleep all along, that life as it could be (and as it is meant to be) has passed us by, and that only by taking those drugs will we become able to experience life to the full, for the first time ever truly awake, truly *there*. But there is more at stake here than just our possible failure to perform well in our jobs or to experience life to the full. As Allen Buchanan and others have argued, if we don't get a lot smarter very soon, we will not be able to deal effectively with the global and potentially life-destroying problems that we face today, and thus will be unable to prevent a catastrophic downturn of human life, possibly leading to the extinction of the whole human race (Buchanan 2011, 158–161; Persson and Savulescu 2012, 1–11). As a species, we are currently just not smart enough to do anything about it.

In this vein, in a TEDx talk recorded in 2012, the neuroscientist and philosopher Anders Sandberg, research fellow at the Future of Humanity Institute at the University of Oxford and a leading transhumanist, announced, with triumphant humility, that he had just made an important discovery, namely that he is stupid. But not only that. He also discovered that all those bright people at Oxford University who made him feel stupid are actually stupid too. In fact, we all are. Humanity is a very stupid species, with the brightest people quite capable of being outsmarted by a mouse. However, since we are also a very *powerful* species, we have a problem on our hands, because stupidity paired with power is obviously a very dangerous combination. We are, Sandberg claims, like monkeys, prodding with a stick at potentially lethal stuff, with no clue what we are actually doing. Clearly, then, something needs to be done about our stupidity. We need to cognitively enhance ourselves (presumably in order to leave the monkey stage behind us and become, through self-directed evolution, truly human at last—though if we really are like monkeys now, that is, not very bright, won't we botch that up too?). Cognitive enhancement is thus understood and presented, by Sandberg and others, as an urgently needed therapy to a fatal disease, the disease being our inborn stupidity as a species, or once again, the human condition as such, which places us on a direct route to extinction.

Moral Enhancement

Diagnosis: We are all evil (or at least not good enough).

Cure: Increasing our capacity and disposition for empathy, love, and fairness.

But even that may not be enough to save us, because our defects are not merely cognitive. In 2008, the Swedish philosopher Ingmar Persson and the ethicist and director of the Oxford Uehiro Centre for Practical Ethics Julian Savulescu published an article in which they warned (quite surprisingly given that Savulescu used to be one of the most vocal proponents of human enhancement, including cognitive enhancement) against the "perils of cognitive enhancement" and claimed that, to combat or avoid those perils, it is imperative that we (also) enhance the moral character of humanity (Persson and Savulescu 2008). So suddenly, another form of enhancement, *moral* enhancement, was needed as a safeguard against the potential dangers of cognitive enhancement. Persson and Savulescu's argument was quite simple (not to say simple-minded): if we make people smarter, bad people (like, for instance, Islamist terrorists) will also get smarter, and if they do, it will be much easier for them to accomplish their evil goals. In subsequent years, they then developed and refined their argument in a series of articles, culminating in the publication of a book with the telling title *Unfit for the Future: The Need for Moral Enhancement* (2012). By then, the main problem necessitating moral enhancement was (wisely) no longer ultra-smart evil terrorists, but rather the fact that we are constitutionally unable to deal with the problems that we face today—with global terrorism, mass starvation in countries that we euphemistically call developing, environmental destruction, and climate change. Liberal democracy does not help, on the contrary: it makes matters worse because it can only ever allow popular policies, and the restrictions that we would have to impose on ourselves in order to save the planet for future generations and non-human animals are never going to be very popular as long as we are morally so restricted as we are. Thus, we tend to believe that we are morally responsible only for what we actively cause, not what we merely allow to happen. Our altruism is usually limited to people that are nearby (in space and time), and we are emotionally unaffected by large numbers, so we can stomach the starvation of millions easier than the starvation of one person right on our doorstep. So the problem is not that we are not *smart* enough. The problem is that we are not *good* enough. As a consequence, we really are about to mess it up, and before long, it will be too late to do anything about it, so what we need to do, if we can, is improve our moral dispositions and find a way to overcome the deficiencies that are part of our evolved nature. Since traditional ways of moral education have proved to be largely ineffective, the only hope we have to achieve this is through moral *bio*enhancement, that is by reconstructing the human condition to make it more amenable to, or increase the scope of, empathy, fellow-feeling, our sense of fairness and moral obligation, and so on.

But is it really so simple? Unfortunately, while we would probably all agree that if we managed to make people fairer, more just, more considerate, less selfish, less hate-filled, and more sensitive and sympathetic to the plight of others, this consensus doesn't get us very far because justice and fairness are very abstract notions, and empathy and sympathy tend to be rather selective. Thus, two people can both believe in justice, but differ considerably about what justice requires. They can both have a lot of empathy for the victims of injustice, but very much disagree on who is the real victim in a given case. Of course there are people who are entirely indifferent to moral considerations, incapable of caring for anyone or anything but themselves, or driven by antisocial desires so strong that they cannot but comply with them, but those people are not the problem that moral bioenhancement is meant to solve. They are only a problem to the extent that they are not fit to live in a society which is built on mutual respect, or not fit to be lived with. They are a permanent danger to other people precisely because they are *not* like other people. They lack something that people normally have and that they need in order to live together peacefully, to live and let live. With respect to that end, they are clearly defective, and if we think we may have discovered a means to cure them of that defect, then we are well advised to consider using it. But this is therapy, not enhancement. It is about bringing people up to a certain standard of normalcy. Perhaps more importantly, those people are usually not the ones who are intent on destroying whole civilisations. On the contrary, terrorists tend to be infused with moral righteousness. If they weren't so convinced that certain things were morally wrong, or indeed evil, if they weren't affected so much by certain notions of harm, then they wouldn't feel the need to cause so much havoc and destruction. It is not the fact that they have no moral conscience that makes them dangerous. Rather, it is the fact that they have too much of it. Naturally, we would probably say (*we* would say) that they don't have the right kind of morality, that they are misguided. But they are clearly not misguided in the sense that they favour injustice over justice, or selfishness over altruism. In their own way, they care very much about justice and the welfare of other people. They just have a different idea about what is unjust and what is good or bad for people. So if we wanted to enhance them morally in such a way that they no longer pose a danger to the survival of humanity, or at least that of Western civilisation, then it wouldn't be enough to simply make them more moral, or more just, or more altruistic, or more empathic. We would have to find a way to make them *think* about these things more like we do. That would, of course, not qualify as a moral enhancement per se, unless we assume that there is a right way to think about these matters, and that our way of doing so is the right way. In other words, we would have to assume that moral realism is true and that we have privileged access to moral truth, both of which are highly controversial, to say the least.

Yet the moral enhancement project has to deal with an even more fundamental problem. Leaving bioweapon-wielding terrorists aside, how would we morally enhance the average good people that seem incapable of dealing effectively with global problems? We may be tempted to think that the problem here is just a lack of will power, that people know very well what they should do, but somehow cannot bring themselves to actually do it. There are certain strong dispositions that prevent us from doing what we see as right. So by changing those dispositions, we would actually become more autonomous. It would enable us to exercise our own free will. But it might not be as simple as that. Our alleged moral myopia, the fact that we discount the future and events that occur in some distant region of the world that have little impact on our lives, and our fondness for the acts-and-omissions doctrine are not really *obstacles* that prevent us from doing what we know we should do, but rather beliefs or dispositions that *directly affect* our moral convictions. They make us less certain that what we are being asked to do really is what we should do. Are we *really* responsible for what happens at the other end of the earth or in the far future? Just as responsible as for what happens now and nearby? Is there *really* no moral difference between actively causing something to happen and merely failing to prevent it? Perhaps there is a good reason for why many of us have such doubts, and I don't mean that in the sense that it reflects some evolutionary pressure of a distant past that has long ceased to be useful or adaptive. It might still be useful *today*. It might protect us from feeling responsible for too much. It might allow us to take better care of the here and now. It might prevent us from seeing and treating people as exchangeable. Never mind whether this is really so. My point is that it is far from obvious that correcting such alleged shortcomings in our moral psychology would really be a moral enhancement, all things considered.

Yet even if we think it is, we may want to think twice about implementing a moral enhancement research programme. Those who, like Persson and Savulescu, believe in the desirability and possibility of moral bioenhancement tend to make much of the biological roots of our moral dispositions and are pretty confident that we will be able to affect the desired changes by manipulating our brain chemistry. Hormones such as oxytocin, which functions as a neurotransmitter, are believed to be connected to our ability and willingness to trust other people as well as to other pro-social dispositions. It should be obvious, though, that the ability to manipulate our moral dispositions has its own dangers, so the proposed therapy may well prove to be worse than the disease (of a lack of morality) that it was meant to cure. Given the diagnosis, we should not expect moral enhancement technologies to be used for the human good. It is much more likely that they would be used to increase some people's power over others.

Life Extension

Diagnosis: Ageing is a disease, and death the greatest evil.

Cure: Radical life and health extension through genetic engineering, nanotechnology, and other biotechnologies.

Then there is death, of course. Radical life extensionists like Max More (1996), Nick Bostrom (2005), Aubrey de Grey (2007), or John Harris (2007), all believe, and encourage us to believe, that death is "the greatest evil", and ageing, because it inevitably leads to death, the worst disease. Max More even argues that as long as we have to die, it is not possible for us to live a meaningful life. Mortal life is, per definition, meaningless. So once again, the human condition—an essential part of which is our mortality—is decried as a state that is deeply, utterly deficient and unsatisfactory. It is, in Harris's sense, by its very nature, a "harmed condition", and our job is to fight it with the greatest possible determination and urgency. "Death is a malfunction of the human experience", says Zoltan Istan. "It's a reversible error, a transitory cloak of emptiness, a curable disease – a highly curable disease if dealt with properly" (Istvan 2013, 271). Various companies are already working hard on it. In 2013, Google announced a new company, Calico (California Life Company), whose CEO is the biochemist and biotech manager Arthur Levinson. Calico's mission "is to harness advanced technologies to increase our understanding of the biology that controls lifespan" (http://www.calicolabs.com/). They are now collaborating with the Broad Institute of MIT and Harvard and with QB3 (the University of California's Institute for Quantitative Biosciences) to advance the understanding of age-related diseases and longevity. But the goal is not merely theoretical but primarily practical: to combat ageing and save us from the plight of death. Other companies pursue a different, more unusual path to achieve the same objective. Thus, Australian start-up company Humai (Human Resurrection through Artificial Intelligence), with its CEO Josh Bocanegra, aspires "to reinvent the afterlife": "We want to bring you back to life after you die. We're using artificial intelligence and nanotechnology to store data of conversational styles, behavioural patterns, thought processes and information about how your body functions from the inside-out. This data will be coded into multiple sensor technologies, which will be built into an artificial body with the brain of a deceased human" (http://humaitech.com/). If we cannot keep people alive, then we shall do our best to raise them from the dead.[3] This is to be achieved by cryoconserving the brains of the freshly deceased and then constructing an artificial body to house the brain, which can then (it is hoped) be reanimated.

Projects such as these are motivated by a moral imperative that transhumanists believe follows from the fact that death is wrong, something that clearly *ought not to be*. "In the future," writes Collin Duncan (2015), "death simply won't be a choice or a technologically eliminated disease. It will be fundamentally, morally wrong at a very deep level. To allow death will be to enable it, much as we see passively handing someone a gun to self-annihilate as assisted suicide today." Teaching children to believe in an afterlife and accept death will be seen as tantamount to child abuse, which is exactly what it is. A political reorientation is urgently recommended ("The proliferation of anti-AI talk needs to stop. The advances of neo-ludditism upon the technocratic future need to be rejected. This isn't an alternative. This isn't an option. This is an imperative."), and a supposedly purely rational utilitarian reasoning is employed to derive the desired result: "Transhumanism produces an infinite hedonic score at an infinite magnitude higher than the alternative to the greatest number of individuals (death vs non-death)." Therefore, Duncan argues, human life *must* be upheld "at all costs". In pursuit of this goal, the Australian branch of Humanity Plus just (in January 2016) launched a petition "to deem ageing as a disease", which we are encouraged to sign before it is sent to "governments around the world".[4]

Popular culture has been quick to catch up with this view. An internet blogger who calls himself "The Jesus Alien" (2009) informs us: "Mortality is a disease and biological immortality is no longer science fiction, scientists know that we will one day be able to adjust our life spans [...] Now, what if some ancient civilisation already mastered this through some herbal mix through extracting chemicals from atralagus or other herb, or what if these people were travellers from another realm or planet with advanced technological knowledge?" This may sound crazy, which it is, of course, but what is interesting about it is the change in attitude towards our own mortality that it betrays. Death is no longer a given, dying no longer a necessity. We can use herbs (or some other natural substance or process) against the disease of mortality. The possibility of a cure proves the existence of the disease, and that a cure is possible is proved by the fact that others have done it before us: just look at Michael Jackson, whose likeness can be found in images (helpfully provided by the blogger) throughout history, going back to ancient Egypt. So clearly, Michael Jackson must be an immortal alien. And if aliens can do it (with their advanced technology), then we can too. The logic is impeccable. And we are all willing to literally buy into it when we listen to the promises of the consumer society, spearheaded by the cosmetic industry. As it happens, Yves Saint Laurent sells a range of skin products under the brand name "Forever Youth Liberator", which very neatly captures the essence of transhumanism, namely the aspiration to be completely, and always, in control, to be a truly autonomous being.

The common theme, traceable in the discourses surrounding each of those different forms of human enhancement, is that technology will cure us from the disease of being human. Once we are cured, we will no longer be human. We will be post-human. It seems that, nowadays, many of us cannot wait to get there. We can sense the impatience, the desperate longing for the promised cure in the tenacity with which crowds are queuing for the latest electronic gadget or telecommunication device in front of the shops, long before they open their doors, just so that they can be the very first to get and own them. Take, for instance, the iPhone, which plays the role of a symbol of the world-to-come, a symbol of the hoped-for post-human condition when we can live our lives, for the first time ever, disease-free. When Apple launched its new iPhone 5s in 2013, its marketing was supported by a video that showed people using their iPhone in various different situations and for all kinds of purposes, underlining the *empowering* nature of the device ("You're more powerful than you think"). The soundtrack to the video was a song by Jennifer O'Connor, called "When I grow up", which contains the following lyrics: "When I grow up, I'll be the hero/of my story book. I'll start out zero./ [...] When I grow up, I'll be good and strong./I'll create a world where I belong. When I grow up, I'll be who I want to." In combination with the video, those lines strongly suggest that it is *technology* and its potential to enhance all our capacities that will eventually allow us to grow up, and once we have done so, we will finally be able to write our own story (rather than have it written for us by the forces of nature, which include our own treacherous body). We will no longer be bad and weak, as we are now, will no longer have to live in a world where we do not belong, and will no longer be prevented from being exactly who we want to be. This comes very close to the benefits that the transhumanist philosopher Nick Bostrom cited when he set out to explain "why he wanted to be a posthuman when he grew up".

Now it may well be true that the *longing* for all that is as old as humanity itself. Yet there is still something new here. The increasing tendency to view human enhancement as a form of therapy (which is meant to cure us from the human condition) marks a remarkable change in our *normative attitude*. It seems to be more and more common to believe that we *deserve* to be enhanced and that we have the *right* to be. And with good reason: if enhancement really is therapy, then it is not unreasonable at all to believe that we are *entitled* to be enhanced just as we are now entitled to be cured when we are ill. It is what common decency seems to dictate, *if*, that is, the human condition is indeed adequately described as a disease. But is it? Do we really need to be cured of our humanity?

This is obviously not a question that has an objective, factual answer. It is a question about our attitudes to life in general and our human condition in particular, so ultimately, each of us has to answer that question for themselves. What I have tried to do in this chapter, and for that matter, this whole book, is mainly share observations and describe phenomena, and not necessarily provide a proper, compelling argument for or against the enhancement–therapy identity thesis, or more generally transhumanism as a worldview. Arguments, especially when it concerns ethical issues (i.e. what is good and bad, what we should and should not do, what is desirable and what is not), are not as important as it may seem. We need to buy into the *stories* on which they are built in order to be convinced by them, need to, in this particular case, accept the persuasive definition that is used to support the claim that being human *itself* is a disease, a defective state. But this also means that there is no compelling argument for the claim that humanity is in need of a cure. It all depends on whether that particular story resonates with our own outlook, our view of the world, and our own role in it. However, Harris's suggestion that the mere *possibility* of an improvement beyond what is normal and healthy according to current human standards brings about a moral *obligation* to provide said improvement strikes me as outlandish, mainly because possibilities are, by their very nature, *endless*. Whatever abilities we have, there is always some ability that we lack, in respect to which we would then have to consider ourselves as disabled. But as Havi Carel (2013, 82–83) has pointed out, being "unable to be is not an independent or context-free concept. It has to be seen in relation to being able to be. An inability to be is a modification of an ability to be that is lost (…) or an ability that is never achieved viewed against a background of a common capability." Thus, the inability to live for hundreds or thousands of years or forever, the limitations of our cognitive powers, our lack of moral saintliness, and the emotional vulnerability that comes with love are not disabilities. If they were, then we would *always* be disabled and stay disabled, no matter how many additional abilities we would acquire. Every ability is limited in some way. What may appear to be a lack of ability, or a disability, from the perspective of the theoretically possible (which, in practice, comes down to what is *imaginable*), is in fact, in the context of actual human existence, an ability: the ability to live, to love, to think, and to do what we think and feel is right. The transhumanist omnipotender, who possesses all abilities and lacks none, is not a sensible ideal. It is itself *not* a possibility.

Buchanan's argument that we are so badly constructed that we cannot survive without changing our very nature is of course a different matter because it does not rely on questionable ethical claims. Convincing, however, it is not. We have survived so far and have actually proven quite resilient. If Buchanan were right, one should expect that we already perished a long time ago. But we did not, and chances are we will be around for some time even without enhancement. And if for some reason we won't, then it is rather unlikely that an enhancement of our nature would have saved us.

Notes

1. For an overview of this early discussion, see Parens (1998).
2. This conception bears only a superficial similarity to earlier conceptions of humanity's alleged deficiencies or sickness, for instance, Friedrich Nietzsche's and Arnold Gehlen's. When Nietzsche declares the human to be "the sick animal" (das kranke Tier) (1966, II, 862: *On the Genealogy of Morality* III.13), he does not define human nature as such, but seeks to describe a common (but by no means natural) present human condition. The sickness of the human is not inscribed in our nature, but shows itself merely in our (culturally induced) attitude. It is a deviation from the healthy attitude of self-respect, courage, and defiance that we should have and could have if we only chose to. An enhancement of our biological nature is in no way needed to restore our health. Gehlen is a different case, of course, because for Gehlen, it *is* our nature to be the "deficient being" (*Mängelwesen*, Gehlen 1940, passim), but for Gehlen, this does not render us in need of a cure because our very deficiency is the reason why we have developed something that is so much more valuable than nature, namely *culture*. Our deficiencies did not harm us: they provided us with the opportunities to become what we are. They are what has made us *special*.
3. The idea of bringing back the dead through technology is not entirely new. It was already promoted more than a century ago by the Russian proto-transhumanist Nicolai Fedorovich (Burdett 2011, 25–28).
4. http://www.transhumanism.com.au/petition

Bibliography

Bostrom, Nick (2005) "The Fable of the Dragon Tyrant," *Journal of Medical Ethics* 31: 273–277.

Buchanan, Allen (2011) *Beyond Humanity?* Oxford: Oxford University Press.

Burdett, Michael S. (2011) "Contextualizing a Christian Perspective on Transcendence and Human Enhancement", in *Transhumanism and Transcendence*, ed. Ronald Cole-Turner, Washington, DC: Georgetown University Press, 19–35.

Carel, Havi (2013) *Illness. The Cry of the Flesh*, Revised edition, Durham: Acumen.

De Grey, Aubrey, and Michael Rae (2007) *Ending Aging*, New York: Saint Martin's Press.

Duncan, Collin (2015) "The Day no one died: The Moral Imperative of Transhumanism (https://medium.com/@BleachedSleet/the-day-no-one-died-the-moral-imperative-of-transhumanism-5d74d6c3c234#.mi709bp9n)

Earp, Brian D., Olga A. Wudarczyk, Anders Sandberg, and Julian Savulescu (2013) "If I Could Just Stop Loving You: Anti-Love Biotechnology and the Ethics of a Chemical Breakup", *The American Journal of Bioethics* 13/11: 3–17.

Earp, Brian D., Anders Sandberg, and Julian Savulescu (2015) "The Medicalization of Love", *Cambridge Quarterly of Healthcare Ethics* 24/3: 323–336.

Earp, Brian D., Olga A. Wudarczyk, Bennett Foddy, and Julian Savulescu (2016), "Addicted to Love: What Is Love Addiction and When Should It Be Treated?", *Philosophy, Psychiatry, & Psychology*, in press.

Gehlen, Arnold (1940) *Der Mensch. Seine Natur und seine Stellung in der Welt*, Berlin: Junker und Dünnhaupt.

Harris, John (1993) "Is Gene Therapy a Form of Eugenics", *Bioethics* 7/2-3: 178–87.

——— (2007) *Enhancing Evolution. The Ethical Case for Making Better People*, Princeton: Princeton University Press.

Istvan, Zoltan (2013) *The Transhumanist Wager*, Futurity Imagine Media.

More, Max (1996) "Transhumanism. Towards a Futurist Philosophy", www.maxmore.com/transhum.htm.

Nietzsche, Friedrich (1966) *Werke in drei Bänden* (W). Ed. Karl Schlechta, Munich: Hanser Verlag.

Parens, Erik (1998) "Is Better Always Good? The Enhancement Project", in *Enhancing Human Traits: Ethical and Social Implications*, ed Erik Parens, Washington D.C.: Georgetown University Press.
Persson, Ingmar, and Julian Savulescu (2008) "The Perils of Cognitive Enhancement and the Urgent Imperative to Enhance the Moral Character of Humanity", *Journal of Applied Philosophy* 25/3: 162–77.
——— (2012) *Unfit for the Future: The Need for Moral Enhancement*, Oxford: Oxford University Press.
Sandberg, Anders (2012) "Shifting Your Brain". Online video clip. YouTube, 4 Sept. 2012: https://www.youtube.com/watch?v=5eqAXD_Loic
The Jesus Alien (2009) "Michael Jackson is not Dead", The Jesus Alien. WordPress, 12 August 2009: http://www.thejesusalien.com/michael-jackson-is-not-dead-hes-an-ancient-egyptian-princess/
Tooley, Michael (1972) "Abortion and Infanticide", *Philosophy and Public Affairs* 2/1: 37–65.

8
The Unfairness of Nature

It has become quite common among those who welcome the prospect of enhancing human nature to argue that human enhancement is not only morally permissible, but also morally obligatory: that we have a moral duty to enhance ourselves and others, or at least to provide people with opportunities to enhance themselves (e.g. Savulescu 2005; Bostrom 2005; Harris 2007; Persson and Savulescu 2008). In the previous chapter, we discussed the transhumanist idea that the whole human condition is best understood as a disease in need of a cure (or a disability in need of fixing). This idea allows us to think of radical human enhancement as a therapy and, consequently, encourages a sense of entitlement: just as we are owed treatment when we are ill, we are also owed enhancement because we are, ultimately, *always* ill (disabled, or defective), simply by virtue of being human. However, one doesn't have to go that far to reach the desired normative conclusion. The duty to enhance is often believed to follow simply from the fact that our natural talents and abilities are not evenly distributed, that what we can and cannot do is, to a large extent, the result of a "genetic lottery" (also referred to as "natural lottery").[1] In consequence, people's chances in life are, through no fault of their own, hampered by a "genetic inequality", which, being entirely undeserved, is

M. Hauskeller, *Mythologies of Transhumanism*,
DOI 10.1007/978-3-319-39741-2_8

clearly unfair, and because it is, we are morally obligated to do something about it and to, as best we can, "level the playing field". Thus, Julian Savulescu and colleagues once referred to those who "do well at sport as a result of the genetic lottery that happened to deal them a winning hand" as the "genetic elite", arguing that it is unfair that not everybody has the same natural constitution (2004, 667). "Nature," they announced, "is not fair" (2004, 668), which would be uncontroversial if it only meant, as it usually does when we say something like that, that the categories of fairness or unfairness do not *apply* to nature (i.e. that nature can *neither* be fair *nor* unfair) or that nature is not concerned with fairness (but solely with, say, the survival of the fittest, or whatever we think determines the directions that living nature takes). Rather, it is intended to convey the sense that nature is *positively* unfair. It is not just a *descriptive* statement that means to explain how nature works (or does not work), but instead, a *normative* statement that is intended to imply the moral unacceptability of the natural condition (or certain features of it). Based on this normative understanding, enhancement is then recommended as a remedy against nature's unfairness. "By allowing everyone to take performance enhancing drugs, we level the playing field. We remove the effects of genetic inequality. Far from being unfair, allowing performance enhancement promotes equality" (2004, 668).[2]

Yet the unequal distribution of abilities is not the only kind of unfairness that can be utilised to support the claim that we have a moral duty to enhance. While the above argument makes use of a purely comparative and thus *formal* understanding of fairness (cf. Hooker 2005), which merely requires that everybody is treated alike or has the same (opportunities, resources, goods) as everyone else, the moral duty to enhance can also be inferred from a more *substantive* account of fairness that emphasises either: (a) desert or (b) need. So even if everyone had exactly the same opportunities, resources, and goods and was treated in exactly the same way, their situation and treatment might still be unfair if they either do not get what they deserve or do not get what they need in order to live a decent life, or both. Although it is seldom explicitly formulated as an argument, the idea that we, that is, *all* of us, have been treated unfairly *by nature* nonetheless informs much of the pro-enhancement and especially the transhumanist discourse. That idea is at work whenever it is argued that our lives have been blighted by various (natural and hence entirely undeserved) defects such as certain cognitive and emotional limitations and the inevitability of ageing and death. This is seen not only as deeply unsatisfactory, but indeed as unfair, not because there are others out there who have it better, but simply because one shouldn't have to live under such conditions, because we didn't do anything (bad) to deserve it and because we, frankly, would have deserved better. This way of understanding the world and our place in it, which is rather common specifically among transhumanists, is succinctly expressed in Nick Bostrom's claim (which used to grace his personal website,[3] but has now disappeared) that once we have enhanced ourselves to a post-human level, our lives will be "as they should have been all along".

In what follows, I will engage mostly with the argument from *formal* unfairness, that is, the claim that we have a moral obligation to enhance people who are, in regard to their natural constitution, worse off than others *because their being worse off is unfair*. I will concentrate on this argument because it seems to be more widely accepted and because I find it more interesting. However, much of what I am going to say about it is also applicable to the argument from substantive unfairness. Moreover, the argument from formal unfairness will ultimately prove to rely on our tacit acceptance of the argument from substantive unfairness.

Moti Mizrahi (2014) has recently argued that natural inequality presents a new problem of evil for the theist. How can a just God allow this to happen? Answering that question may not appear particularly pressing to many of us, but what should be of interest even for those of us who are not theists is the assumption that Mizrahi makes and that indeed must be made in order to support his claim that there is a problem in the first place. That assumption, which Mizrahi treats as obviously true, is that natural inequality is morally obnoxious, and that the reason for that is its unfairness: "It is important to note that natural inequality is an evil, not because it leads to pain and suffering, although it might and often does, but because it is unfair" (Mizrahi 2014, 130). Since the talented deserve their talent as little as the untalented deserve their lack of talent, it is, "from a moral point of view", "not fair that one person is taller, healthier, faster, thinner, more intelligent, more beautiful, more agile, and otherwise more naturally endowed than another person". Note that the intended argument does not seem to be that this situation is unfair because God doesn't do anything about it although he could, but rather that God should do something about it because the undeserved inequality of abilities and other desirable traits is indeed unfair.

The argument from formal unfairness clearly draws on the presumed value of equality and raises the question as to what extent we (or God, but that is a different discussion) are morally obligated to promote it. Natural differences between people obviously exist, that is, differences that are not owed to their own choices and actions, but to their genetic and epigenetic endowment.[4] Inequality is thus undeniably a basic fact of life. Yet recognising this fact does not compel us to hold that all undeserved inequality is inherently unfair and needs to be redressed. This is because despite this view's obvious intuitive appeal, it is not entirely clear whether it actually makes sense to declare a natural condition, which is not in any way the result of human or any other being's agency, to be "unfair". Is it unfair that I cannot bend it like Beckham, or play golf like Tiger Woods? We would not normally think so. I may well find it unfair that those people earn so much more money than I, or that they are famous and I am not, based on my conviction that I deserve fortune and fame just as much as they do. But it would be odd to claim that I deserve their *talent* as much as they do. The reason why I might find it unfair that I am not as rich and famous as they are is that my own abilities, even though they are different from theirs, may conceivably appear to me as worthy of being rewarded and publicly recognised as theirs. In other words, my sense of being treated unfairly rests on my conviction that I do in fact possess certain noteworthy, reward-deserving abilities myself. Those abilities then form the *normative foundation* of my claim that I am being treated unfairly. Yet if I *lack* certain abilities, then I also lack the abilities that I would need to support my claim that I am just as deserving of those abilities as those who actually have them. Because the problem is precisely that I do *not* have them, and that there are no abilities that I have *prior* to my existence that could possibly serve as a normative foundation for my claim that I deserve certain abilities that others are born with but I am not. I can certainly *begrudge* others their talent, but that is not the same as finding it unfair that they have it and I do not. It is perhaps psychologically not impossible to do so and we might even feel tempted to rationalise our grudge and frustration by calling the unequal distribution of talents "unfair", but it is definitely a stretch,[5] and we know it.

Leaving aside the question of desert, the difference between an unequal distribution of fame and fortune, on the one hand, and of ability, on the other, and the reason why it makes more sense to us to call the former unfair, is that fame and fortune are ultimately bestowed on certain people by human agents, while ability is not. They only get rich and famous because we appreciate their abilities and are willing to reward them. We pay good money to see them play, and we take a keen interest in their lives. If we didn't, they wouldn't be rich and famous at all. So in a way, we are *giving* them their fame and fortune and at the same time *withholding* it from others. But their abilities are nothing we have given to them. In fact, they have not been given to them by anyone, and that is why it sounds strange to say that it is unfair that they have those abilities and I do not. It would indeed make much more sense if we thought that there is a God or some other agent in the world who is responsible for the distribution of abilities, who actually made a choice and decided, for apparently no good reason at all (because we didn't exist before that decision was made and therefore couldn't possibly have deserved it), that they should have a particular ability and I should not have it. So by claiming that the unequal distribution of abilities is unfair, we actually seem to imply that there is such an agency in the world, that is, that we have in fact been *treated* badly by whomever or whatever is responsible for our lack of talent, in making the *choice* to withhold it from us.

This is why the metaphor of the "genetic lottery" seems peculiarly inapt to convey the view that natural differences are unfair. This is so because lotteries are *not* usually considered to be unfair at all. We all know that if you win a lottery, then this is not because you have done anything to deserve it. In fact, the outcome is completely independent of anything you might have done or not done (except taking part in the lottery) or anything you might be or not be. That is why you can be happy about it, but not proud. It is completely arbitrary, you were just lucky. You could just as well have not won. This is considered not unfair, but on the contrary, entirely fair. It would only be considered unfair if the odds were stacked, that is, if your win in the lottery were no longer arbitrary but due to someone doing something that made it more likely that you won the lottery rather than somebody else. In other words, lotteries are considered fair precisely to the extent that their outcomes *are* arbitrary, that is, determined by chance, and unfair to the extent that their outcome is *not* arbitrary, that is, not determined by chance. We can then infer that the unequal distribution of abilities is perfectly fair, not *despite* being determined by chance and thus the result of a "genetic lottery" but *precisely* for that reason.

Now it could be objected that it is not, and was never meant to be, the unequal distribution of natural assets *itself* that is inherently unfair. Rather, what is, or would be, unfair is not to redress this unequal distribution once it becomes possible to do so. Thus, while it is not unfair that some people are less intelligent than others, even if it does restrict their choices in life, it would be unfair to allow this inequality to persist. That is the position that John Rawls suggested in his *Theory of Justice*: "The natural distribution is neither just nor unjust; nor is it unjust that persons are born into society at some particular position. These are simply natural facts. What is just and unjust is the way that institutions deal with these facts" (Rawls 1999, 87). Yet if we accept that we should try to, as much as possible, guarantee fair equality of opportunity, which requires not merely that all legal obstacles to acquiring certain desirable positions in life be removed (careers open to talents), but also that everyone have the same chance to attain those positions (Rawls 1999, 63), then it seems that we should also try to distribute the natural goods (such as health and vigour, intelligence, and imagination) equally since they, without doubt, influence our chances in life just as much as, if not more than, the social conditions we are born into. "There is no more reason to permit the distribution of income and wealth to be settled by the distribution of natural assets than by historical and social fortune" (Rawls 1999, 64). Since those natural inequalities are undeserved, that is, morally arbitrary, we should, according to Rawls, definitely try to redress them by, for instance, spending more money and effort on the education of those who are less intelligent (Rawls 1999, 86). Only very briefly, however, does Rawls entertain the idea that we may try to change nature itself, and instead of merely compensating for a lack of natural abilities, make sure that everyone has the abilities they need to pursue a preferred plan of life. For although he admits that it is indeed "in the interest of each to have greater natural assets", he is unwilling to infer more from this than that society should "take steps at least to preserve the general level of natural abilities and to prevent the diffusion of serious defects" (Rawls 1999, 92), which is still a long way from demanding that natural abilities be distributed equally and ultimately not consistent with the "level playing field concept of equal opportunity" (Buchanan et al. 2000, 65) that Rawls seems to have endorsed.

This inconsistency was addressed and partly corrected by Allen Buchanan and colleagues in their seminal and aptly named treatment of the issue, *From Chance to Choice*: "If precise and safe control over the distribution of natural assets becomes feasible, then those who believe that justice is concerned with the effects of natural assets on individuals' life prospects will no longer be able to assume that justice requires only that we compensate for bad luck in the natural lottery by intervening in the social lottery, rather than by attacking natural inequalities directly" (Buchanan et al. 2000, 64). If the goal is to create a situation where everybody starts life with the same opportunities, then it doesn't seem to matter whether what stands in the way of that goal is a certain social structure or an uneven distribution of natural assets. In fact, once we have learned to manipulate our nature, the distinction between social assets and natural assets becomes blurred, if not altogether untenable (Parker 2012, 128). The same holds for the distinction between social inequalities and natural inequalities (Lewens 2009). It seems, therefore, that there is indeed a case to be made for the removal of natural inequalities (Holtug 1999, 139),[6] and some have indeed made this case very forcefully. Mark Walker, for instance, has argued that since people who are naturally very happy (hyperthymic), because they are genetically predisposed to have a higher average level of happiness than most people, tend to have much better opportunities in life, we have a moral duty to provide those who were less fortunate in the genetic lottery with the means to acquire the same high level of average happiness. That duty is derived from the alleged unfairness of the fact that not everybody is as happy as the hyperthymic. "To deny the rest of us access to HPP (Happy People Pills) is a grave form of injustice, for it is to artificially limit the pool of this valuable resource; and to restrict the pool of this valuable resource is to prohibit most of us from the opportunity for what many (but not all) of us see as the best life: life with the happiness and achievement of the hyperthymic" (Walker 2009, 35).

However, if those who are naturally happy or hyperthymic did nothing to deserve such happiness, then it is hard to see why we should think that those who are *not* naturally happy deserve to be happy. It is true that we do not deserve to be unhappy. But we do not *deserve* to be happy either. Yet if we neither deserve to be unhappy nor deserve to be happy, then it seems that considerations of fairness simply do not apply. It would only be unfair to leave people in their (after all only comparatively) unhappy constitution if it were unfair for them to *be* in that condition in the first place. Thus, the argument from formal unfairness presupposes some degree of substantive unfairness. There are two ways in which the claim that it is *substantively* unfair that not everyone is by nature extraordinarily happy could be plausibly defended. It could be unfair if: (a) we all *deserved* to be extraordinarily happy, or (b) we *needed* to be extraordinarily happy to live a decent life, that is, one from which none of the basic goods are missing. Yet since nobody needs to be hyperthymic to live a good life and there is no good reason to suppose that we deserve to be better than we actually are, the argument from substantive unfairness also appears unconvincing.

But let us assume for the moment that we accept the above argument in favour of enhancement for the sake of equal opportunity. If, say, A is by nature more intelligent or more happy than B so that A has a better chance of getting on in life, we should either compensate B for their undeserved handicap in the game of life, or lift B, provided that is possible, up to the intelligence or happiness level of A so that both have the same starting conditions. Those who show themselves to be persuaded by this kind of argument usually suppose A and B to be humans. But is there any good reason why we shouldn't apply the underlying principle to humans and non-humans alike? Consider the following argument: I cannot fly, but if I could fly I would have many opportunities that are not open to me now. It would also give me great pleasure. Thus, the quality of my life would arguably be enhanced if I could fly. Yet there are other living beings that can fly, for instance, birds. This is a result of the genetic lottery which has given birds abilities that have been denied to me. That is unfair because the fact that birds can fly and I cannot is morally arbitrary. Birds did nothing to deserve their ability to fly, as I have done nothing to deserve my inability to fly. Therefore, society has a moral obligation to provide me with the means to acquire that ability, should such means be either already available or at least in our reach. Since there are many different species of animals, most of which have abilities that humans do not currently possess, there is obviously a lot to enhance before we will actually have levelled the playing field completely. Bats can echolocate. Dogs have a sense of smell that is 10,000 times more acute than ours. Ants can carry loads that exceed their own body weight multiple times. And so on and so forth. Is all this unfair? Are we morally entitled to be able to do what they can do? I suspect that most of those who believe that we have a moral duty to level the playing field by enhancing people who seem to have lost out in the genetic lottery would not want to go that far. But why should we regard the species boundaries as relevant here? Why should the genetic lottery suddenly stop to be unfair (and thus in need of correction) when we cross the species barrier?

The argument works of course in both directions. Not only are there many things that animals can do that we cannot do, but there are also things that *we* can do and that animals cannot do. So should we also level the playing field for them? Nick Bostrom (2004) may have been the first who suggested as much in his whimsical fantasy about Albert, the uploaded and cognitively enhanced golden retriever who is interviewed on the Larry King show. After reflecting on the unequal distribution of luck among both humans and animals, the former dog, now post-dog, Albert declares himself to be the "lucky one", and when asked by Larry King whether he had a mission in life, he replies that he wants "everyone to be the lucky one". Given the context, this can only mean that he wishes for all animals to be subjected to the same cognitive enhancement process that he himself has undergone. That animal enhancement or "uplifting", which we already discussed in Chap. 6, is indeed a matter of distributive justice has been expressly argued by George Dvorsky (2008). Dvorsky claims that we have a duty to cognitively enhance animals because not to do so "would be an unfair distribution of primary goods that are requisites for political participation, liberty and justice". This is thought to be unfair because it is the result of pure luck: "Like some humans who argue that they have fared poorly in the genetic lottery, it can be said that nonhumans have missed out in the species lottery" (2008, 136). Others, like Sarah Chan (2009), have also argued that we have a moral obligation to enhance animals, although not directly on the grounds that we would be treating them unfairly if we didn't. Rather, the claim is based on the fact that animals have interests and that they matter morally just as much as ours. But it seems to me that this argument ultimately also appeals to our sense of fairness: the reason why we should not only take human interests into account but also the interests of animals is, presumably, that an unequal treatment is not morally justifiable. In other words, it would be unfair to take heed of human interests but disregard the interests of animals.

Yet if we do level the playing field so thoroughly that everyone (be they human, animal, or machine) has all the abilities that everyone else has, so we have ended the tyranny of the genetic lottery, then all differences would have disappeared. It would be a thoroughly equalised (*gleichgeschaltet*) world. Why would we want to live in such a world? Differences clearly also have their value. They make life interesting and rich. They give us an individual identity. While a certain degree of equality may be desirable, universal sameness certainly is not. Fortunately, it is also highly unlikely, if only because it may be impossible to have *all* the abilities that can currently be found in anyone. Some of those abilities are likely to be mutually exclusive, so you cannot have the one *and* the other. Levelling the playing field completely would then not only be undesirable but also impossible.

Now, the same holds if we restrict our ambitions to the human world. I cannot bend it like Beckham or play golf like Tiger Woods, no matter how hard I try. I just don't have the ability. Nor can I play the violin like Yehudi Menuhin, or paint like Picasso, no matter how hard I try. This is no fault of my own and therefore morally arbitrary. But it is hard to see how we should be able to have all those different abilities combined in one and the same person. Should we then try to level the playing field to the extent that it is possible?

The problem with this suggestion is that before we can start attacking natural inequalities, we first need to clarify what exactly we intend to make equal. Arguably, it is chances in life, but it is not always obvious which natural assets increase our chances in life, and how much of any of those assets we need to possess to have better life prospects. One should think that being smart or good-looking is not exactly a disadvantage in the game of life, but there is no evidence that it is the smartest people who are economically most successful, or, even more importantly, who are most likely to have the best or happiest life (Veenhoven and Choi 2012). And how smart exactly do we have to be? It is certainly not the case that the smarter you are, the more opportunities in life you will have and the better your life is likely to be. Whether a so-called natural asset is indeed an asset for you depends, to a large extent, on the circumstances, not only those created by the reigning social structure, but also on what other abilities you have and what kind of opportunities emerge in the course of your life. "Thus, there is no such thing as a resource per se. Different traits will be resources in different social environments. Recognition of this simple fact imposes a fundamental constraint on any attempt to intervene in the natural lottery in the name of equality of opportunity or resources" (Buchanan et al. 2000, 80).

In some cases, it is even difficult to know whether having an ability or not having it makes for better opportunities in life, and hence, whether it is unfair that you have it or unfair that you do not have it. If, for instance, you have few or no moral scruples and are prone to stealing stuff and killing people, you are likely to end up in prison, which is clearly not fair since you have never asked to be bad in the first place. It is just your bad luck of being born with the wrong genes and into the wrong social circumstances. Hence, since you haven't really chosen to be bad,[7] we owe it to you to compensate you or, if we have a way to do that, to change your nature in such a way that you are no longer bad. We have a moral obligation to make you good. But then again, we could just as well argue the opposite. A lack of moral scruples, or let us say a certain moral flexibility, is not always a disadvantage in life. It is, after all, often people with moral integrity who miss out on opportunities that may make their life better, at least in the sense that it may give them access to certain goods and resources and increase their liberty. So it seems that we owe it to you to make you less good, so that you are not disadvantaged in the game of life by your moral scruples.

This shows two major problems with the whole argument from formal unfairness. The first is that equality of opportunity may not be the only relevant value at stake here. Perhaps it is better for people to have moral integrity even though it does diminish their ability to acquire certain goods. The other problem is that we don't always know what exactly enables us to have a good life, and that is the fundamental problem of the whole enhancement project (Hauskeller 2013). If it would not necessarily and obviously be better for me to be as intelligent as those who are more intelligent than I, then there is no clear reason to think that the situation is unfair. In order for the fact that you have something that I do not have to be considered unfair, what you have must (as a *necessary* condition) be, in some sense, better than what I have. It must be considered beneficial or, in some other way, worth having, but whether or not it is worth having depends on many contingencies as well as on our own individual idea of a good life.

It is of course not entirely impossible to identify abilities that appear generally beneficial, but this is easiest and least controversial in the case of abilities whose lack seriously limits our options in life. If I am positively stupid or hideous to look at or constantly depressed, I may have a problem, but this has got nothing or little to do with the fact that you are more intelligent or better-looking or generally more upbeat than I. The problem is not comparative in nature. If I am merely less intelligent or less handsome than you, I may not be able to reach the same professional heights as you or to find a partner who is physically as attractive as yours, but if my intelligence and my looks don't fall too much under the average, then there is no reason to assume that I cannot also have a rich, fulfilling, and professionally successful life, and that is all I need and all I can reasonably want from life. What matters is not how much smarter you are (in comparison with me), but merely how smart I am and whether I am smart enough to fit in and to be able to live a good life. The same holds for all other natural assets. It is good to have some of them or all of them to some reasonable degree, but I don't have to be the best in everything to live a good life. Although with my limited abilities there are a lot of things in life that I cannot do, or cannot achieve, it would still be impossible to do most of those things even if I possessed all possible abilities to the highest possible degree. True, with my natural assets, I could never hope to be a David Beckham or a Tiger Woods or many others who excel in a particular area, but I could never be all of them anyway, especially not if everyone else would also be a natural born Beckham or Woods.[8] As long as there is something I can do, something I am good at, and nothing that prevents me from fully participating in the world and society in which I live, I should be just fine. Why should we all want to be good at everything anyway? That is why Buchanan and colleagues very wisely concluded that "at least for the foreseeable future (if not forever), the appropriate objective (…) may be something more like the attainment of a 'genetic decent minimum' (…) than the elimination of all inequalities in natural assets" (Buchanan et al. 2000, 82). With this, I can whole-heartedly agree.

The idea that we should all have exactly the same starting conditions to guarantee that we also have more or less the same opportunities in life strikes me as wrong-headed from the outset. It frames life as a competition and then seeks to eliminate all differences that might give anyone any advantage over anyone else, which makes any meaningful competition impossible.[9] We do compete with other humans, and we do compete more with them than we compete with other non-human species. That may be the reason we tend to ignore other animals and their abilities when we consider what is fair and what is not. But there is nothing really to be gained by levelling the genetic playing field for all humans so that nobody starts the competition with a disadvantage. Equality, even equality of opportunity, does not have any obvious intrinsic value. In fact, it does fairly obviously *not* have any intrinsic value. If we made the smarter ones dumber, the happier ones less happy, and the prettier ones less pretty, then we would do as much for equality of opportunity as when we did the opposite. But nobody suggests that, and with good reason.[10] Life is not a race, and ethicists should not have to suppose that it is. We don't necessarily compete with each other. And even if we did (or where we do), it couldn't work if we were all the same. If the resources are limited, then we can only achieve equality by holding people back. If we don't want to do that, if we do accept inequality at some stage, then we will, by guaranteeing the equality of the starting conditions, just postpone the time that inequality will materialise, which seems to defeat the purpose of the whole enterprise.

Now, it is commonly assumed that while it would be unfair not to grant everyone the same starting conditions, it is *not* unfair that people achieve different things in life through their personal virtues: their determination and dedication, the effort and work they put in to accomplish something, and so on. But of course all this also requires certain abilities that are as much based in our genetic code as our intelligence, our longevity, or our propensity to develop cancer.[11] If it is unfair that I am less intelligent than you, then it is also unfair that I am lazier than you, less committed, less dependable, less trustworthy, and so on. So we would have to equalise these abilities too in order to *really* level the playing field, and once we have done that, the only differences that would remain, if any, would be down to brute luck. So ironically, by levelling the playing field completely, in order to annihilate morally arbitrary distinctions, we would create a situation in which *all* differences would become morally arbitrary. That is because if there still are differences in people's situation, they could no longer result from differences in intelligence, talent, or virtue, since all those differences have already been eradicated. Our whole life path would then be a lottery: entirely fair and at the same time entirely arbitrary.

Notes

1. Clearly, the term "natural lottery", which John Rawls, in his *Theory of Justice*, first published in 1971, seems to have introduced into the debate on social and political justice, and what exactly such justice consists in and what it requires, is somewhat broader than the term "genetic lottery". The natural lottery encompasses all differences between people that are *not* due to social circumstances or human action, but have come about through natural processes. Whether those differences have been caused by differences between people's *genetic* constitution or by other factors that influence our development before birth is irrelevant. In contrast, the term "genetic lottery" emphasises the presumed *cause* of those differences and, in doing so, already suggests the possibility of a solution to the moral problem whose existence the term is meant to express—namely, the *unfairness* of there being such differences, or of not doing anything about those differ-

ences. The solution that is suggested is of course some kind of genetic intervention. I will, in this chapter, mostly use the term "genetic lottery" because it better conveys the sense that it is indeed possible not only to counterbalance the *effects* of that lottery, but also to actually tackle and change the lottery itself, namely so that it is no longer a lottery. However, since I do not here attach any importance to the *causes* of existing natural differences, genetic or otherwise, I will use the term "genetic lottery" generally in the sense of "natural lottery".

2. It is of course highly doubtful that we would do sport a favour by making athletes more equal. As Lisa S. Parker has pointed out (Parker 2012, 128), it is the goal of athletic competition "to establish the inequality of the competitors". Moreover, the "presumption of inequalities is integral to sport's entertainment value".
3. www.nickbostrom.com
4. It has been pointed out to me that claims that we need to redress the genetic lottery presuppose genetic determinism. However, I don't think that is necessary. I'm pretty sure I could have trained as hard as I liked, but I would still never have become a star athlete. That doesn't mean of course that if I *had* had the right genes, then I *would* have become one. It just means that not everyone is born with the same abilities and the same potential to develop certain abilities. Not everyone can be turned (trained or educated) or can turn (train or educate) themselves into anything they want. There are limits to what we can achieve in life that are rooted in our individual nature. And that is all, I think, that we need to assume here.
5. If we agree that the unfairness claim presupposes some kind of desert (so that if I don't deserve X, then it cannot possibly be unfair that I don't get X), and that we cannot deserve anything before we are something, because whatever we deserve, we deserve on the basis of what we are or do, then it seems to follow that it cannot be unfair that I don't have the same natural abilities as you. Thus, the argument is logically valid. Whether we also consider it sound depends of course on whether we find the premises persuasive.
6. Torbjörn Tännsjö (2009, 325) has even suggested that the reluctance to support and promote enhancement for those less favoured by nature is owed to a "Nietzschean view of justice, according to which it is unfair if those who are less fit pool their resources and rob the genuinely strong Übermensch of his genetic advantage". Needless to say, Tännsjö recom-

mends that we get rid of this anti-democratic and indeed "fascistoid" notion of justice as quickly as possible.

7. Peter van Inwagen (2001) has argued that even if somebody has a genetic disposition to a certain kind of antisocial or criminal behaviour (for instance, rape), this fact does not diminish their responsibility for committing the crime as long as they could have acted otherwise. In other words, only if the behaviour in question is genetically *determined* (rather than just genetically *influenced*) should we regard this as a mitigating circumstance. However, even if van Inwagen is right, my point here is not about moral responsibility, but about the fairness or unfairness of having a *propensity* to a certain kind of behaviour. Even though it might be considered more unfair that I am bad and you are not if I just *can't help* being bad, it is arguably still unfair if being good is just *much harder* for me than for you.
8. And if everyone were, then we would certainly not value their talents much anymore. What we do value is the exceptional, not the commonplace.
9. It would also, as Allen Buchanan (1995, 113) has pointed out, expand the domain of moral responsibility to tyrannical dimensions.
10. It might be thought that with regard to social assets, we do in fact occasionally pursue a levelling-down approach to equal opportunity, for instance, when we implement progressive taxation systems that requires the rich to contribute a higher percentage of their income than that which the poor are being asked to pay. However, the reason for this is not that we want to create equal opportunities or make sure that nobody is disadvantaged (and nobody advantaged), but simply that we assume, with good reason, that the rich can *afford* to pay a higher percentage of their income than the poor. We don't want to take away their advantage. We just think, in accordance with Rawls' difference principle, that everyone should benefit from their having more than the rest.
11. This does not mean, of course, that all our abilities and talents are to an equal extent determined by our genes. Rather, my point is that there is *no* ability that is fully mine in the sense that it doesn't rely for its existence on certain aspects of my given nature, which I did not choose and which I did nothing to deserve.

Bibliography

Bostrom, Nick (2004) "Golden", www.nickbostrom.com.

——— (2005) "The Fable of the Dragon Tyrant," *Journal of Medical Ethics* 31: 273–277.

Buchanan, Allen (1995), "Equal Opportunity and Genetic Intervention", *Social Philosophy and Policy Foundation* 12/2: 105–135.

Buchanan, Allen, Dan W. Brocks, Norman Daniels, and Daniel Wikler (2000) *From Chance to Choice. Genetics & Justice*, Cambridge University Press.

Chan, S (2009) "Should we enhance animals?", *Journal of Medical Ethics* 35: 678–683.

Dvorsky, George (2008) "All Together Now: Developmental and ethical considerations for biologically uplifting nonhuman animals", *Journal of Evolution and Technology* 18/1: 129–140.

Harris, John (2007) *Enhancing Evolution. The Ethical Case for Making Better People*, Princeton: Princeton University Press.

Hauskeller, Michael (2013) *Better Humans? Understanding the Enhancement Project*, London: Routledge.

Holtug, Nils (1999) "Does Justice Require Genetic Enhancements?", *Journal of Medical Ethics* 25: 137–143.

Hooker, Brad (2005) "Fairness", *Ethical Theory and Moral Practice* 8: 329–352.

Lewens, Tim (2009) "What Are 'Natural Inequalities'?" *The Philosophical Quarterly* 60/239: 264–285.

Mizrahi, Moti (2014) "The Problem of Natural Inequality: A New Problem of Evil", *Philosophia* 42: 127–136.

Parker, Lisa (2012) "In Sport and Social Justice, Is Genetic Enhancement a Game Changer?", *Health Care Analysis* 20: 328–346.

Persson, Ingmar, and Julian Savulescu (2008) "The Perils of Cognitive Enhancement and the Urgent Imperative to Enhance the Moral Character of Humanity", *Journal of Applied Philosophy* 25/3: 162–77.

Rawls, John (1999) *A Theory of Justice. Revised Edition*, Cambridge, MA: Belknap Press.

Savulescu, Julian, Bennett Foddy, and Megan Clayton (2004) "Why We Should Allow Performance Enhancing Drugs in Sport", *British Journal of Sports Medicine* 38: 666–670.

Savulescu, Julian (2005) "New Breeds of Humans: the Moral Obligation to Enhance", *Reproductive BioMedicine Online* 10, Supplement 1: 36–39.

Tännsjö, Torbjörn (2009) "Medical Enhancement and the Ethos of Elite Sport", in *Human Enhancement*, eds. Julian Savulescu and Nick Bostrom, Oxford: Oxford University Press, 315–326.

Van Inwagen, Peter (2001) "Genes, Statistics, and Desert", in *Genes and Criminal Behavior*, eds. David Wasserman and Robert Wachbroit, Cambridge: Cambridge University Press.

Veenhoven, Ruut, and Yowon Choi (2012) "Does Intelligence Boost Happiness?", *International Journal of Happiness and Development* 1/1: 5–27.

Walker, M.A. (2009) "The Case for Happy-People Pills", *Free Inquiry* 29/5: 33–36.

9

Gods Rather than Cyborgs

In this chapter, we will look at the ambivalent relation that transhumanism has to religion, and particularly to the idea of God. I said earlier that transhumanists, unlike Donna Haraway, would, if they had to choose, prefer to be gods rather than cyborgs. But this is more than just a preference. It is a moral mission, based on a deep-rooted sense of injustice: the perceived unfairness of being merely human, of finding ourselves with desires and ambitions and dreams that we will never be able to fulfil. There is a painful gap between, on the one hand, what we can imagine ourselves to be and can therefore *dream* of being and, on the other, what we can ever, as long as we are tied to the human condition, *hope* to be. Our human nature allows us to see and long for the stars, but prevents us from reaching them. So if there is a God, he is to blame for our limitations, our collective disability, for all that we cannot do and be; at the same time, he, or his image in our minds, is a constant reminder of what one *could* be and do. God, for mainstream transhumanism, is the being that, if it exists, has chosen to dangle the fruits of all-encompassing knowledge and eternal life right before our eyes and noses, but strictly forbids us (and if needs be, prevents us by force) to eat them, lest we become (as *Genesis* 3.22 reminds us) as powerful as he is. A god who does that is a cruel god,

M. Hauskeller, *Mythologies of Transhumanism*,
DOI 10.1007/978-3-319-39741-2_9

a jealous and taunting god, one who does not deserve to be revered, but needs to be fought and dethroned instead. We resent the king who does not deign to let us eat at his table, who thinks we are not his equals, and, like the proud angel Lucifer, the bringer of light, we vow to right this wrong by bringing him down and putting ourselves in his place. Most transhumanists, of course, do not believe in God. However, the transhumanist's professed lack of belief is often so fervent, so pronounced, that it seems grounded in a deliberate *refusal* rather than a mere failure to believe, almost as if they wanted to punish God for his unfair treatment of us by denying his existence. But that denial does not make the resentment go away. Rather, the resentment we may harbour towards an unjust God, transhumanists commonly transfer (in a Freudian sense) to religion as an institutionalised belief in God or some other higher power.

Many transhumanists strongly emphasise the non-religious nature of their views, and are openly and programmatically opposed to religious faith of any kind. Not only does the transhumanist, relying on reason alone, not need any religious faith, but religion is also viewed and represented as a hostile force that needs to be combated and defeated before we can hope to realise our human aspirations to transcend our current limitations. Religious faith does not only happen to be an obstacle to progress. In the standard transhumanist mythology, religion's main *function* is to hold us back and to keep us on our knees (More 1996). It is, in this respect, a real-world representative of the biblical God who jealously guards his own privileges. "God was a primitive notion invented by primitive people, people only just beginning to step out of ignorance and unconsciousness. God was an oppressive concept, a more powerful being than we, but made in the image of our crude self-conceptions. Our own process of endless expansion into higher forms should and will replace this religious idea" (More 1996). In Istvan's *The Transhumanist Wager* (2013), all opposition to transhumanism is blamed on religious fanatics, who bang on about "devil's work", Satan, and hell fire, and would love to see all advocates of scientific and technological progress "burn at the stake"—as if no other, perhaps more rational motivations could possibly be behind anyone's reluctance to wholeheartedly pursue the transhumanist political agenda. If you are against it, you must be corrupted by the unholy influence of religion and the "deathist" culture it has created. That is why, without exception, all characters in Istvan's novel who have got anything to do with religion are portrayed as corrupt, hypocritical, superstitious, technophobic, power-hungry, and essentially evil. Yet as Hopkins (2005) has pointed out, transhumanism and religion have more in common with each other than with secular humanism. They both seek transcendence and liberation from a perceived unsatisfactory mere-animal condition of the human. They both promise salvation from the burden of our earthly woes and our mortality. Transhumanists may not believe in sin, but they do tend to see the human condition as a punishment (though certainly not a just one). They even have their own eschatology, most clearly in the concept of the "singularity", which is the point in time (expected very soon) when the machines we build will have become so incredibly smart and powerful that all bets are off and literally *anything* becomes possible. When[1] that happens, then the world as we know it will end. For this reason, the singularity has been aptly described as the "rapture of the geeks" (DeBaets 2015). Where transhumanists differ most from those adhering to any of the traditional monotheistic religions—apart from the obvious: their refusal to accept any power greater than their own—is with regard to the chosen *means* of achieving said transcendence of, and salvation from, the current human condition. Transhumanists favour, as the preferred, if not only viable means, science and technology, in whose salutary power, however, they put just as much trust and faith as any Christian can put in his God. Transhumanism is therefore indeed, as Tirosh-Samuelson (2012) has pointed out, a "secularist faith" in the sense that it constitutes itself as a hybrid between the religious and the secular, which as such is no longer entirely secular, but has rather become, by re-enchanting the secular, "post-secular".

It should not come as a surprise, then, that there are various fringe transhumanist groups, some of which seek to establish transhumanism as an entirely new religion, while others try to integrate transhumanism into traditional religious belief systems. An example of the former is *Prometheism*, which presents itself as the "21st century Religion of Transhumanism", which is supposed to be founded not on myth and superstition, but on reason and science, and is dedicated to, unsurprisingly, Prometheus. To join the religion, you are asked to swear an oath to help create a "future of boundless possibilities" and to promote "self-directed evolution", and particularly the creation of "a neo-eugenically enhanced race that will eventually become a new, superior species".[2] Prometheism also happens to be openly racist (which most other transhumanist organisations are *not*). Other post-secular transhumanist organisations are more conciliatory, seeking to merge with, rather than replace, traditional religion. The Mormon Transhumanist Association, based in Utah (where else?), advocates a religious transhumanism called "transfigurism", which promotes "radical flourishing" through technological change and the progressive *theosis* (deification) of humanity.[3] The Hollywood-based Church of Perpetual Life, which is inspired by the writings of the aforementioned Russian visionary Nikolai Fedorov and whose main goal is radical life extension, wants to "accelerate the Creator's plan of the Common Task of Humanity, which is to cultivate technology that will facilitate the transformation of life into an environment of perpetual duration".[4] So-called blind faith is to be replaced by the supposedly more rational faith in science and technology, which, it is believed, will eventually bring us the immortal lives that (an apparently no longer oppressive) God wants us to have. In an interview (Cuthbertson 2015), Neil VanDeRee, a minister of the Church of Perpetual Life, confidently declares: "I'm not afraid of death because I know I'm not going to die", adding, "It might sound crazy, but some people believe in angels in heaven and fiery hell. For them, that's their truth. Immortality is mine." And even though he does not believe in heaven, he says, he does believe "that the earth should be the utopic heaven that many Christians envision in the sky after they're dead. I believe that our task as a species has the choice of returning this planet to a utopic garden and we should inhabit that place forever. (…) I believe we should all be gods. I believe once we obtain immortality for our species that we will be gods because we'll be able to be immortal. We will have unlimited lifespans and the sorts of technologies that will look like magic to people living today. We are destined to be gods."

The belief that humans are *destined* to evolve further and to eventually become gods, that we are not just here by accident, but that our lives have an objective purpose, that we have a crucial role to play in a divine or, at any rate, cosmic plan that is ultimately directed towards our own deification is a fundamentally religious belief that is shared by many transhumanists. The Terasem Movement, founded by Martine Rothblatt, which describes itself as a "transreligion for technological times" and keeps "mind files" of its members for the purpose of future mind uploading and mind cloning (which involves the creation of multiple selves), lists as their first "core belief" the belief that "life is purposeful" and that its purpose is "to create diversity, unity and joyful immortality everywhere". The other core beliefs are that "death is optional", that "love is essential", that "God is technological", which means that God is not simply there, but is in the process of becoming, or more precisely in the process of being made (by us) through the implementation of "technology that is ever more all-knowing, ever-present, all-powerful and beneficent".[5] Gabriel Rothblatt, Martine's son, who is also a "pastor" with Terasem, states: "The end goal of Terasem is similar to other religions, these ideas of joyful immortality in the afterlife. But for us it's not simply a spiritual concept, it's a mechanical challenge. Technology could one day make this a reality through digital backups – the idea of transferring a person's consciousness on to a hardrive, which could then be placed into quasi-utopian conditions. Heaven could be a virtual reality world hosted on a computer server somewhere." (Cuthbertson 2015). Even those who have already died without having had the foresight or the means to create a mind file of themselves may eventually be able to enjoy this technology-enabled heaven. As some transhumanists speculate, everything that has ever happened is likely to be permanently stored in "a cosmic memory field hidden in yet unknown aspects of reality". If that is the case, then we just have to figure out how to access this field (which we of course will someday), and then we shall be able to technologically resurrect everybody who has ever lived (Prisco 2015).

Not all transhumanists may harbour such fantasies, but what transhumanists who identify their views as in some way religious have in common with those who are openly opposed to all things religious (which is the majority) is that they all aspire to transcend the human condition and to "transfigure" humans into gods. "Life," says Istvan (2013, 69) "is essentially a choice between pursuing personal godhood or dust." And he is right, of course. Only a god is immune from the fate that we share with everything there is: that one day, our existence will inevitably come to an end. Only gods are immortal. So it is either death or godhood, which for the transhumanist is a no-brainer: dust is no option; so godhood it shall be. However, whether we really have a *choice* in the matter, which is here simply assumed, namely as a corollary of a supposed human destiny, is far from clear. The structure of the universe may not allow for the existence of gods. We may still give it a try, though, either by creating machines with godlike powers who serve us or, preferably (because it is difficult to see why any entity with godlike powers should *want* to serve us), by acquiring or transferring such powers to ourselves. "Imagine," writes the prophet of the singularity, Vernor Vinge (1993), "a willing slave, who has 1000 times your capabilities in every way. Imagine a creature who could satisfy your every safe wish (whatever that means) and still have 99.9 % of its time free for other activities. There would be a new universe we never really understood, but filled with benevolent gods (though one of *my* wishes might be to become one of them)." Since filling the universe with benevolent machine-gods willing to grant all our wishes (or, in other words and rather paradoxically, with all-powerful slaves) is a rather risky endeavour—the genie, once out of the bottle, may well decide that he does not want to go back in again—it is a lot safer to become a god oneself, thus ensuring slavedom for everyone else, because, logically, there can be only one all-powerful being. This is the paradox of autonomy (which will be the main focus of the next and last chapter of this book).

Some transhumanists are actually so keen to achieve a god-like status and so convinced that human self-transformation and apotheosis is nothing less than a divinely ordained mission that they are willing to throw all caution overboard. In their recent book *The Proactionary Imperative* (2014), the sociologists Steve Fuller and Veronika Lipinska wish to persuade their readers that we should no longer let our policies and legislation on scientific research and technology development be guided by the *precautionary principle*. Instead, we should adopt the *proactionary principle* (formerly advocated by Max More).[6] While the precautionary principle demands that we refrain from conducting risky scientific or social experiments that we cannot be reasonably sure will not endanger human lives or well-being, the proactionary principle encourages us to take even considerable risks if the potential benefits appear sufficiently great. Fuller and Lipinska believe that the precautionary principle should be discarded because it decelerates scientific and technological progress, which they think is bad *not* because, as others (such as More) have argued, we need to have that progress to enable us to successfully deal with the problems that we face today (such as global warming) or because without it we are stuck with all sorts of deficiencies and evils (such as the inevitability of ageing and death). Rather, throwing obstacles in the way of progress is bad simply because it *goes against our human destiny*. Humans, we learn, are meant to be more than just clever animals, because in contrast to all other living beings, we are "touched by the divine", have "a 'god-like' character" (Fuller and Lipinska 2014, 1), and are "central to the cosmos" (8). Being human is all about self-transcendence, about gradually turning ourselves into the gods that we potentially are and that we have always been meant to be. Since the precautionary principle makes it difficult to realise our divine potential, it needs to go.

Although *The Proactionary Imperative* is full of references to the classics of our intellectual history, so one would be hard-pressed to name a single prominent theorist that is not being mentioned at least once, the book's scholarly packaging cannot hide the fact that the only argument we are being offered in support of the recommended 180-degree turn in public policy is based on nothing more than a religious conviction. Thus, the "foundation for transhumanism" that Fuller and Lipinska promise to give us is the completely unfounded proposition that moving ever forward and becoming the masters of the universe is, well, what we are really here for. In their view, self-transcendence is more than just an entitlement: it is a sacred duty. All the other, rather unorthodox policy propositions follow from this one article of faith: that we should introduce "legal arrangements that would encourage people to invest themselves or their capital in risky scientific experiments" (4), not worry about "the freedom of future generations" (9) or overpopulation (43), "conceptualize our genetic material as property that one is entitled, and perhaps even obliged, to dispose of as inherited capital" (33), curtail the civic rights of anyone who refuses to participate in (potentially dangerous) scientific research (38), welcome the "species culls and large-scale environmental restructurings" we have brought about in the past (82), endorse "mass surveillance and experimentation, with the understanding that many in retrospect may turn out to have been used or sacrificed for science" (63), and "remove criminal sanctions from the conduct of risky experiments" (111).

It is of course to be expected that casting all caution to the wind will cost us. Lots of innocent people are likely to be harmed. Many will die. But that is fine, Fuller and Lipinska inform us, because that is a price we must pay—and each of us should be more than happy to pay—for the advancement of humanity. We need more "experiments in living, regardless of outcomes" (43). We should of course compensate the victims, but how exactly remains unclear. Perhaps we can be re-educated in such a way that we will all feel already sufficiently compensated by being honoured as heroes, the martyrs of scientific progress. However, Fuller and Lipinska make it clear that compensation is not really necessary because we actually *owe* it to society (and humanity) to risk our lives for scientific progress. We have, after all, been "allowed to live" (107), and now it is time to give something in return. In addition, by willingly turning ourselves into means to a higher end, we give a meaning to our lives that they would otherwise lack (99).

Strangely, for Fuller and Lipinska, the greatest danger that we face today is not that, because we are too eager to test "the limits of what is possible" (26), we may inadvertently destroy the very grounds of our existence or a life still worth living, but that we may "sleepwalk into a suboptimal future" (36). To avoid being suboptimal, we are being encouraged to risk everything. But is that really enough to make us turn our backs on the precautionary principle? That one can only support the precautionary principle if one presumes "that 'Nature' sets a non-negotiable norm to which we and other living beings must ultimately conform" (37) is of course nonsense. One does not have to presume any such thing. What one *does* have to presume is merely that the life and well-being of the individual matter, and that no religious or crypto-religious belief in humanity's ultimate destination should prompt us to disregard them. In contrast to what Fuller and Lipinska would have us believe, the end does *not* justify the means and we are *not* "entitled to adopt…God's point of view" (132). What they embrace, and ask their readers to embrace, is the kind of ruthless optimism that Schopenhauer once accused Leibniz of: one that is not worried by any evil that might result from our actions because of the firm conviction that everything is for the best and cannot but turn out just fine. If one does not share that religiously motivated conviction, then the proactionary principle has little to recommend it.

And if Fuller and Lipinska are right that the transhumanist movement rests on the plausibility of the proactionary principle, then we can safely conclude that the whole philosophy of transhumanism lacks a coherent (secular and rational) foundation. "But what else could possibly justify transhumanism other than the literal belief in our own capacities for apotheosis?" (45) What indeed!

But it is not only the belief that we are *destined* for godhood that motivates the transhumanist demands that we do whatever it takes to get us there. There is an unmistakable sense of urgency in the transhumanist project that cannot be fully explained by a belief in some cosmic plan that involves our eventual rise to godhood. If that were all, then there would be no reason to hurry. We could just let things gradually unfold until we have become what we have always meant to be. What makes our rise to godhood so *urgent* that we have got no time to lose is the conviction that the kind of life that is open to us now is not only worse than it could conceivably be and worse than it shall be in future, but that a mortal life, filled with all kinds of pain and suffering and overshadowed by the certainty of a never-far-off death, is not really worth much at all. Thus, Max More (1996) has argued that death undercuts meaning, in the sense that as long as our lives will have to end someday, our lives cannot be truly meaningful. Even religion with its promise of a life after death can only ever achieve the illusion of meaning, but never the real thing. This is mainly because true meaningfulness cannot be derived from being part of somebody else's (in this case God's) plan, which has the inevitable effect of stifling a sense of our own personal value. Yet without such a sense of personal value, More claims, our (individual) lives must lack true meaning, because what gives our lives (true) meaning is "the continuation of the process of improvement and transformation of ourselves into ever higher forms". Since this process is understood as open-ended, it is clear that death, by bringing it to an end, destroys not only the meaning that any individual life can have up to the point of its termination, but the very *possibility* of meaning. If our lives can only have meaning if we can pursue "our own expansion and progress without end", then life can only be meaningful if it *never* ends. That certainly plays a part in why death is often perceived and described by transhumanists as the greatest evil (Bostrom 2005; de Grey/Rae 2007, 36). It is the greatest of evils not merely because it sets an end to our life, to our aspirations, hopes, and plans (cf. Nagel 1979; Nussbaum 1994, 207–210), or because it deprives us of all future pleasures (Bradley 2004), but because it renders all we do *meaningless*.[7] For this reason, the argument goes, we need to do everything in our power to stop the presently inevitable decline of our bodies and find a way to extend human lifespan indefinitely. It would follow that ultimately only an immortal, God-like existence can meet the minimal requirements of a meaningful life and, since it is difficult to see how a life entirely devoid of meaning could qualify as good, also the minimal requirements of a (minimally) *good* life. So in other words, life is only worth living for an immortal god, and as long as we are not such gods, life is not worth living for us. In fact, if we have no choice but to live our lives as mere humans, it might be better for us not to have been born at all.

This is the point where transhumanist anti-deathism connects with anti-natalism, as advocated principally by the South African philosopher David Benatar. In his book *Better Never to Have Been: The Harm of Coming into Existence* (2006), Benatar skilfully defends the seemingly absurd view that we would all be better off if we had never been born and that, precisely for this reason, it is morally *wrong* to bring children into existence and not to abort a foetus before it comes into existence "in the morally relevant sense at around twenty-eight or thirty weeks gestation" (Benatar 2006, 148), and morally *desirable* that our species (and indeed all sentient species) go extinct earlier rather than later. On the face of it, that does not sound like a position that a transhumanist is likely to adopt. Transhumanists, after all, tend to attach considerable value to life and their own continued existence. They most certainly do not believe that non-existence is generally preferable to existence. Notwithstanding this apparent incompatibility between anti-natalism and transhumanism, I shall, in the remainder of this chapter, argue that the spirit of anti-natalism is in fact congenial to the transhumanist understanding of life and the human condition.

Benatar argues that even if one's children are going to have a comparatively good life (which one can never be sure of in advance), it is still never good enough to outweigh the harm of existence, and the longer humanity carries on with prolonging its existence by procreation, the more unjustifiable suffering there will be. According to Benatar, non-existence (or more precisely not coming into existence, which is different from ceasing to exist) is always preferable to existence. This is so for the following reasons: first, even the most blissful human life is still subject to various forms of inevitable suffering: "pain, disappointment, anxiety, grief, and death" (2006, 29). No matter how lucky you are, it is simply not possible to avoid all of these harms once you have started existing. The only way to avoid them is by not coming into existence. "Only existers suffer harm" (ibid.). Second (and most crucially), this suffering is not outweighed by the many good things that you may enjoy when you are alive, even if those good things in your life by far outnumber the bad things. While this may be sufficient to make your existence worth *continuing*, it is not sufficient for your life to be worth *starting*. The good things cannot outweigh the bad things because there is a basic asymmetry between pleasures (positive experiences, satisfied preferences, or goods of any kind) and pain (negative experiences, unsatisfied preferences, or the lack of goods) such that the absence of pain is good even if that good is not experienced by anyone, while the absence of pleasure is not bad unless that absence is experienced by someone. So in other words, while non-existence is better than a bad existence, it is not worse than a good existence. This asymmetry explains why we tend to believe that it is a moral duty not to bring people into existence that we know are likely to have a miserable life, but not that it is a moral duty to bring people into existence that are likely to have a (comparatively) good life. If we wanted to insist on the symmetry between pleasure and pain, then we would either have to claim that there is nothing wrong with bringing people into the world that we know will have a miserable life, or that we are morally obligated to bring as many happy people into the world as possible. If we are not prepared to subscribe to either of those two views, then we have to accept the asymmetry between pleasure and pain. Yet if it is good to prevent the existence of a life with pain in it, but not bad to prevent the existence of a life with pleasure in it, then it follows that even "a life filled with good and containing only the most minute quantity of bad – a life of utter bliss adulterated only by the pain of a single pin-prick – is worse than no life at all." (2006, 48).

Benatar knows very well that few people will be willing to accept his conclusion, no matter how compelling his argument may be. The world is, after all, full of "cheery optimists" (2006, 211) who stubbornly and against all logic cling to the belief that their life is, all things considered, not so bad (and much better than it actually is), that bringing children into the world is a good thing or at least not something that is generally morally wrong, and that we have a moral obligation not to endanger the continued existence of humanity. However, as Benatar argues, these deeply ingrained intuitions are not trustworthy because they are simply the psychological effect of evolutionary pressures. We only think that way because it promotes the survival of the species: "Those with pro-natal views are more likely to pass on their genes." (2006, 8). That is why we are very good at seeing the silver lining, but not so good at seeing the cloud, whose continued existence we tend to ignore. Instead of seeing life as it really is (namely "a piece of shit when you look at it", to quote not Benatar, but Monty Python), we are "engaged in a mass self-deception about how wonderful things are for us" (2006, 100). The fact that most people do not regret having come into existence does therefore not count against the argument because it is not rational reflection that leads people to be happy with their existence, but their "primal" psychological biases, which have been shaped by the process of natural evolution.

Now since I am a cheery optimist myself (since I do not regret having come into existence and do not feel guilty of having brought others into this world), I find it difficult to agree with Benatar's conclusion. However, I am happy to accept that while we do not have a moral duty to cause the existence of happy people, we do have a moral duty not to cause the existence of unhappy people. Not causing the existence of happy people is not wrong, but causing the existence of unhappy people is. It is also hard to deny that we would not be worse off if we had never existed. What I do *not* agree with, however, is the claim that we would have been *better off* if we had never existed. While existence may not be preferable to non-existence, even if that existence is rich and rewarding, neither is non-existence generally preferable to existence (though it might be in some cases). If that is correct, then we do not have a duty to procreate (at least not for the sake of those we bring into existence), but neither do we have a duty *not* to procreate. Benatar's claim that non-existence is preferable to even the best possible human existence gains its plausibility not so much from the asymmetry claim (which can easily be defeated),[8] but from the evolutionary debunking argument that suggests we vastly overestimate the quality of our lives. But for this to be even possible, we need to assume that we may be mistaken in finding our lives worth living. What Benatar is saying is that even though we may be perfectly happy with our lives, we *ought* not to be happy, that even though we may not regret at all having been brought into existence, we *ought* to regret it. Life is, in fact, pretty bad, but we are constitutionally unable to see it. Yet if we don't perceive our lives as bad, how can they be *in fact* bad? We may want to say that there are certain features that a human life must have in order to be called good. But normally we would seek to establish a list of such objective good-making features by looking at what actual lives we think go well. But this, Benatar cannot do because he believes that there are no such lives. What he does instead is postulate a counterfactual state of complete autonomy as the norm for a good life, which, incidentally, feeds into the transhumanist narrative that the current state of humanity is fundamentally deficient and, in comparison with what is theoretically possible, a harmed state, or a state of disability: "Paraplegics may require special access to public transport, but the inability of everybody to fly or to cover long distances at great speed means that even those who can use their legs require transportation aids. Our lives surely go less well for being so dependent. Our lives also go less well because we are susceptible to hunger and thirst (that is unable to go without food or water), heat and cold, and so on. In other words, even if disability is socially constructed, the inabilities and other unfortunate features that characterize human lives are enough to make our lives go very badly – indeed much worse than we usually recognize" (2006, 119). In other words, our lives are bad because we lack complete independence, because we need stuff and because it is not fully under our control whether we get what we need. Transhumanists would wholeheartedly agree with this assessment. However, it is far from obvious that our neediness is something that makes our lives on the whole bad (and worse than if we weren't needy creatures). More importantly, it is not even more *realistic* to regard our various dependencies in that way. It is not in any way closer to the truth of the matter. It simply betrays a different attitude to life and what makes it good. Transhumanists, however, can adopt Benatar's view and argue that as long as we don't radically enhance ourselves so that we are no longer dependent on food and water, temperature, and transportation aids, we would be better off dead, so that the only justification for continuing our existence as a species is a determined effort to pursue a transhumanist agenda of overcoming all our dependencies. It all fits together perfectly: the transhumanist dissatisfaction with the current human condition and Benatar's "pro-death view".

It seems that Benatar has now realised that himself. In his most recent book (Benatar and Wasserman 2015), he expresses his sympathy with the transhumanist worldview (thereby lending further support to my previous observation that we are currently experiencing a "transhumanisation" of our culture, which includes our intellectual culture). Writing in support of his claim that our lives are actually much worse than we commonly recognise, he now says: "The sad truth, however, is that, on the spectrum from no knowledge and no understanding to omniscience, even the cleverest, best educated humans are much closer to the unfortunate end of the spectrum. There are billions more things we do not know or understand than we do know and understand. If knowledge really is a good thing and we have so little of it, our lives are not going very well in this regard. Similarly, we consider longevity to be a good thing (at least if the life is above a minimum quality threshold). Yet even the longest human lives are fleeting. If we think that longevity is a good thing then a life of a thousand years (in full vigour) would be much better than a life of eighty or ninety years (especially where the last few decades are years of decline and decrepitude). Ninety is much closer to one than it is to a thousand. It is even more distant from two or three or more thousand. If, all things being equal, longer lives are better than shorter ones, human lives do not fare well at all." (Benatar and Wasserman 2015, 52). Yet the comparative (compared with what, though?) shortness of our lives is not the only aspect of our human condition that allegedly makes our lives bad: "Human lives would (…) be immensely better if we lived for many thousands of years in good health and if we were much wiser, cleverer, and morally better than we are" (53). Because being human is in all respects comparatively bad (meaning that for every apparent good that we enjoy, we can always imagine something that is not only better, but a *lot* better), it does not make any sense to want to hold on to the present human condition. "To prefer a human life to a better life suggests a distracting sentimentality about humanity. It is to think that it is more important to be human than to have a better quality of life" (58). Fortunately, however, "not all optimists fetishize humanity. Among the advocates of human enhancement are those who envisage and welcome the project of a 'post-human' future – a future in which humans have been so enhanced (physically, mentally, and morally) that they are no longer recognizably human. These advocates of transhumanism think it is much more important to improve the quality of life than for the enhanced future beings to be human. While there are many who object to the wisdom and morality of seeking such enhancements, I am not among those categorically opposed to technological enhancements" (60). So there we have it. If life is not worth living as it is, then we can only avoid the conclusion that it will always be better for us not to have been born at all (and hence that we owe it to future generations of humans to make sure that they never come into existence) by ensuring a post-human existence that is infinitely better than anything any mere human can ever hope to have. Where Benatar thinks he disagrees with transhumanism is mostly with regard to the optimistic expectation that with radical enhancements our lives will soon be good enough. Enhanced post-human lives will certainly be much better, but that doesn't necessarily mean that they will be good. Why? Because "we would still die and we would still have vastly more ignorance than knowledge" (62). But this is of course a misunderstanding of transhumanism. For the true transhumanist, our lives will never be good enough until we no longer have to die at all, we know everything that there is to know, and can do anything and everything that there is to do. The assumption that we are *always* going to die and that we will *always* have vastly more ignorance and knowledge is precisely what transhumanists do not accept, and *cannot* accept if they want to stay true to their beliefs. So the transhumanist response to Benatar's assessment has to be that while his pessimistic view regarding the current human condition is perfectly justified, he is overly pessimistic with regard to the post-human future that awaits us. Just wait, have faith, I imagine them saying, we are already working on it. Immortality and omniscience might be difficult to achieve, but eventually, if we try hard enough, we will figure it all out. And then, when we are for all intents and purposes like gods, even Benatar will have to agree that our lives are now, finally, worth living.

Notes

1. Characteristically, the singularity is not regarded as a possibility, but as something that is *bound* to happen. The only thing that is uncertain about it is *when* it will happen.
2. http://www.prometheism.net/
3. http://transfigurism.org/pages/about/
4. http://www.churchofperpetuallife.org/#!about/c55t
5. The Truths of Terasem: http://terasemfaith.net/beliefs/
6. For More, the precautionary principle enforces "the tyranny of safety" (More 2013, 261), and by focusing on possible future risks rather than actual present dangers created by technological stagnation, it does more harm than good. Instead of continuing to "wallow in a culture of fear" (264), we should focus on the opportunities that technology provides, embrace the proactionary principle as our guide, and thereby protect "the imperative to progress" (267).
7. In contrast, Leon Kass (2002, 2003) has argued (as did a younger Nussbaum 1994, 225–232) that it is mortality that makes life matter in the first place. It is the knowledge that we will have to die, and not too far in the future, that makes things and people precious to us, that inspires love and a sense of beauty and the good, and that is ultimately the source of human dignity or self-worth. And without self-worth, a meaningful life is not possible.
8. Generally speaking, it is neither wrong (morally prohibited) nor right (morally required) to reproduce. Instead, it is, in the absence of particular circumstances that speak against or for reproduction here and now, merely permissible. If we decide to reproduce, then there will be new people whose existence we have caused, and we do have moral obligations to them, just as we have moral obligations to any other person that exists or is going to exist. If we know that the child that we are going to have is likely to have a bad life, full of suffering, then we may justly be accused of having caused harm to that child by allowing it to have such a bad existence. Likewise, if we know that the child we are going to have is likely to have a life that lacks most of the pleasures, satisfied desires, or benefits that we feel constitutes a good life, then we may also be accused of having caused harm to that child. In both

cases, the child that we would have harmed is not the possible child, but the actual child that will exist if we decide to bring it into the world. The presence of pain (or more generally any kind of harm) and the absence of pleasure (or any kind of benefit) are equally bad for that child and may serve as a good (moral) reason for us to abstain from procreation. It is thus not the case that while the absence of pain is good, the absence of pleasure is merely not bad. The one is as bad as the other (especially as the absence of goods is generally a cause of suffering). While neither harms a possible child, both harm any actual child that may result from our decision to procreate.

Bibliography

Benatar, David (2006) *Better Never to Have Been. The Harm of Coming into Existence*, Oxford: Oxford University Press.

Benatar, David, and David Wasserman (2015) *Debating Procreation. Is It Wrong to Reproduce*, Oxford: Oxford University Press.Benater and Wasserman (2015) has been changed to Benatar and Wasserman (2015).

Bostrom, Nick (2005) "The Fable of the Dragon Tyrant," *Journal of Medical Ethics* 31: 273–277.

Bradley, Ben (2004) "When Is Death Bad for the One Who Dies?", *Nous* 38/1: 1–28.

Cuthbertson, Anthony (2015) "Virtual reality heaven: How technology is redefining death and the afterlife", International Business Times, 9 December 2015: http://www.ibtimes.co.uk/virtual-reality-heaven-how-technology-redefining-afterlife-1532429

DeBaets, Amy Michelle (2015) "Rapture of the Geeks: Singularitarianism, Feminism, and the Yearning for Transcendence", in *Religion and Transhumanism. The Unknown Future of Human Enhancement*, eds. Calvin Mercer and Tracy J. Trothen, Santa Barbara: Praeger, 181–197.

De Grey, Aubrey, and Michael Rae (2007) *Ending Aging*, New York: Saint Martin's Press.

Fuller, Steve, and Veronika Lipinska (2014) *The Proactionary Imperative. A Foundation for Transhumanism*, Basingstoke: Palgrave Macmillan.

Hopkins, Patrick D. (2005) "Transcending the Animal. How Transhumanism and Religion Are and Are Not Alike", *Journal of Evolution and Technology* 14/2: 13–28.

Istvan, Zoltan (2013) *The Transhumanist Wager*, Futurity Imagine Media.

Kass, Leon (2002) "L'Chaim and its Limits: Why not Immortality?", in: Leon Kass, *Life, Liberty and the Defence of Dignity: The Challenge for Bioethics*, San Francisco: Encounter, 257–274.

——— (2003) "Ageless bodies, happy souls", *New Atlantis* 1/1: 9–28.

More, Max (1996) "Transhumanism. Towards a Futurist Philosophy", www.maxmore.com/transhum.htm.

——— (2013) "The Proactionary Principle. Optimizing Technological Outcomes", in: *The Transhumanist Reader*, eds. Max More and Natasha Vita-More, Chichester: Wiley-Blackwell, 258–267.

Nagel, Thomas (1979) "Death", in Thomas Nagel, *Mortal Questions*, New York: Cambridge University Press, 1–10.

Nussbaum, Martha (1994) *The Therapy of Desire: Theory and Practice in Hellenistic Ethics*, Updated Edition 2009, Princeton: Princeton University Press.

Prisco, Giulio (2015) "The Quest for Akashic Physics", *Turing Church*, 28 November 2015: http://turingchurch.com/2015/11/28/the-quest-for-akashic-physics/#more-3603

Tirosh-Samuelson, Hava (2012) "Transhumanism as a Secularist Faith", *Zygon* 47/4: 710–734.

Vinge, Vernor (1993) "The Coming Technological Singularity", https://www-rohan.sdsu.edu/faculty/vinge/misc/singularity.html.

10

Automatic Sweethearts

One of the "truths of Terasem" that we briefly looked at in the previous chapter is that "love is essential". But do gods love? If love is a human disease, a condition that is yet another source of potentially great suffering and one that makes us, to some degree, dependent on the object of our love, then we cannot be truly autonomous as long as we have not got rid of love. Gods, then, do not love. They are self-sufficient and do not need anyone else. If they did, they would not be gods (yet). They may have sex, though, because sex is a source of intense pleasure, and pleasure, for the transhumanist, is a prime good. I have explored the transhumanist understanding of sex and love in my book *Sex and the Posthuman Condition* (Hauskeller 2014). What I am going to do here, in this last chapter, is discuss three questions that I believe I have not sufficiently addressed there and that shed light on how the transhumanist ideal of reaching complete autonomy and acquiring godlike powers affects the relations we have with other people. The discussion will focus on the transhumanist desire to replace real human lovers with artificial ones. If we assume, as several futurists profess to believe (Kurzweil 1999, 142–148; Levy 2008, 22; Pew Research Center 2014, 19), that within a few decades, we will be able to build robots that do all the things that we would normally expect

M. Hauskeller, *Mythologies of Transhumanism*,
DOI 10.1007/978-3-319-39741-2_10

a real human lover and sexual companion to do, and that do them just as well, we need to know whether they will then also be, as lovers and companions, as satisfying as a real person would—or whether we will have reason to think or feel that something is amiss, that they are, in some way, not as good. This is our first question. To answer it, I shall assume that those robots will *not* be real persons, by which I mean that although they may give the *impression* of being a person, they are *in fact* not persons. A person, as I am using the term here, is a being that is both self-aware and self-concerned. A being is self-aware if there is (to use Nagel's felicitous phrase) something it is *like* to be that being, and it is self-concerned if it *matters* to it what happens in the world, and especially what happens *to it*. A *real* person is a being that does not merely *appear* to be self-aware and self-concerned, by showing the kind of behaviour that we have learned to expect from a self-aware and self-concerned being, but also one that really *is* self-aware and self-concerned. A being that only *behaves* as if it were a person, without being one, I shall call a *pseudo-person*.

However, in initially making the assumption that those robotic sexual companions of the future will not be real persons in the specified sense, I am not committing myself to the view that it will never be possible for us to create artificial persons. While I do not think that this is very likely, I am happy to concede that, since we do not know what exactly gives rise to self-awareness and self-concern, we can at this stage not entirely rule out the possibility that one day we will be able to create machines that *are* real persons. *If* that happened, then those robots would either be designed to reliably perform certain tasks, say to love, cherish, obey, and sexually gratify us, or they would not. If they were *not* designed to reliably perform such tasks, and instead, were free to make up their own mind about what they want to do and what not (to the same extent that we are), then we would have little if any reason to create them in the first place (except perhaps to see whether it is possible to do so), simply because they would not in any relevant way differ from human persons (which are, after all, much cheaper to produce). It is therefore most likely that if we figure out how to create self-aware and self-concerned robots, we will also seek to make sure that they always do what we want them to do and nothing else, or, preferably (to avoid certain ethical issues, which will be briefly addressed later on), that they always *want* to do what we want them to do. This leads me to my *second* question: would an artificial person (a real one, not a pseudo-person) who has been designed and programmed to reliably give us exactly what we expect a human lover to give us, namely both the actions and the accompanying emotions, thoughts, and attitudes, be, as a lover and companion, as satisfying as a person is (be they human or human-made) who gives us all this *without* having been designed and programmed to do so?

However, although what we experience as satisfying and what not, to some extent, depends on *what* we are (namely as human beings with certain instincts and needs that we all share), it also depends on *who* we are (namely as individuals with certain personalities, attitudes, and world views that may well differ from those of others). For this reason, what satisfies me may not satisfy you, and vice versa. So the two questions raised above—namely whether pseudo-persons would, as lovers and companions, be as satisfying as real persons, and whether real persons who are free (in the sense of not being programmed to obey inbuilt commands) would be as satisfying as real or pseudo-persons who are not free—should not be understood as questions about *actual* levels of satisfaction, but rather as questions about possible *grounds* for satisfaction and dissatisfaction. We will see, though, that those grounds can appeal differently to different people, such that the very same feature that makes an object or relationship appear more satisfactory to some people can make it appear less satisfactory to others. Thus, what we may regard as a vital *defect* in pseudo-persons, one that would make them less satisfactory to us than real persons and thus give us grounds to reject them as adequate lovers and companions, we may also see as an *asset*, something that actually makes them superior to real persons. This consideration gives rise to my *third* and last question: on what grounds can sexual companion robots be regarded as being not only just as good as human lovers, but in fact as better, that is, as *more* satisfactory.

In a footnote to his book *The Meaning of Truth* (1909, 189), William James briefly considers whether an artificial lover could pragmatically ever be as satisfying as a real human one. He imagines this artificial lover, which he calls "automatic sweetheart", as a "soulless body which should be absolutely indistinguishable from a spiritually animated maiden, laughing, talking, blushing, nursing us, and performing all feminine offices as tactfully and sweetly as if a soul were in her". By "soul", James of course means subjectivity or a first-person perspective: an inner life that accompanies and motivates those loving and caring actions that he describes and discreetly alludes to as "feminine offices". The automatic sweetheart would do all those things that we expect them to do exactly like they would if they really felt what their actions suggest they feel, that is, if they really loved us and really cared for us. Except that they do not. Ex hypothesi, an automatic sweetheart does not feel or think anything. They are not real persons, but merely pseudo-persons. As mindless service providers, they would simply perform certain functions, and perform them perfectly. Would that be enough? Would that give us all we need and want? James is certain that it would not, for the following reason: "Because, framed as we are, our egoism craves above all things inward sympathy and recognition, love and admiration. The outward treatment is valued mainly as an expression, as a manifestation of the accompanying consciousness believed in." So what James is saying here is that what we *value* in others (due to the way we are "framed", i.e. to our human nature), or, at any rate, what we value in those with whom we have an intimate relationship, is not primarily the fact that they behave or treat us in a certain, seemingly loving way, but that they do so precisely *because they love us*.

However, it is difficult to see what this love (the subjective feelings and thoughts of which the behaviour is supposedly a mere expression) should consist in, if not in a certain kind of loving *behaviour*. If my lover treats me badly and does not show any concern for my well-being (by, say, looking after me when I'm sick, or by taking care of my needs), then it does not seem to make much sense to insist that they, despite all, do love us. And vice versa, if their behaviour towards us is unfailingly caring and loving and respectful of our needs, then we would not really know what to make of the claim that they do not really love us at all, but only *appear* to do so. We would expect that the alleged lack of love would *show* in some way, and if it never does, then their love is as real as it can possibly be. The philosophical behaviourist Edgar Arthur Singer raised this objection against James in his book *Mind as Behavior* (1924, 9). While a "soulless sweetheart" is indeed unsatisfactory, he argued, their soullessness does not consist in the absence of feeling, but in their behaviour, in what they do and do not do:

> [N]o one would regard a soulless sweetheart as a full equivalent for a soulful one, as these words "soulless" and "soulful" are ordinarily used. But just there is the point: how are they ordinarily used? If I imagine myself come to believe that my mistress, with all her loveliness, is really without soul, I cannot think what I should mean by this if it be not that I fear her future conduct will not bear out my expectations regarding her. Some trait or gesture, a mere tightening of the lips, hardening of the eye, stifling of a yawn, one of those things we say are rather felt than seen, would have raised in my mind the suspicion that she might not to my fuller experience of her remain indistinguishable from a spiritually minded maiden.

On this view, we do not, in fact, *infer* the presence of (a certain kind of) mind from a person's behaviour. Rather, their behaviour *is* their mind (Singer 1924, 10). This is exactly the position that David Levy adopts in his *Love & Sex with Robots* (2008, 11): "There are those who doubt that we can reasonably ascribe feelings to robots, but if a robot *behaves* as though it has feelings, can we reasonably argue that it does not? If a robot's artificial emotions prompt it to say things such as 'I love you,' surely we should be willing to accept these statements at face value, provided that the robot's other behaviour patterns back them up."

Yet the reason why it may not make much sense to doubt the love of somebody who unfailingly behaves lovingly towards us is that we would be hard-pressed to come up with a plausible explanation for why they would do such a thing. By far, the best explanation for their loving behaviour is that they really love us. This does not show that there is no clear distinction between real love and loving behaviour (or more precisely a behaviour that is, *qua* behaviour, indistinguishable from a behaviour that is inspired by real love). How we feel is one thing, and how we behave is, despite obvious connections between the two, quite another. While we usually, though by no means necessarily, express our feelings and attitudes through our behaviour, so our behaviour is normally a reliable indicator of how we feel, we can also hide our "soul" and act as if we felt very differently. Moreover, we know from self-experience that we are beings whose actions are more than just movements in physical space. Instead, they are *always* interwoven with, and expressive of, self-awareness and self-concern. We are real persons, and we *know* that we are. We also know that whatever a person does, there is *some* connection to the subjective side of their existence. A robot, however, is a machine primarily designed to behave in a certain way and, depending on its purpose, perhaps also to make us *believe* that there is something it is like to be that robot. Those companies that today are already producing and marketing social robots (including sexual companion robots) do their best to blur the difference between real persons and pseudo-persons and encourage us to get emotionally involved with their products. This strategy seems to be paying off. As Matthias Scheutz has pointed out, we are hardwired to ascribe intentions to entities that are mobile and exhibit some degree of autonomy, and thus easily fall prey to the "suggestive force of apparent autonomous behaviour" (Scheutz 2012, 213), and it is likely that the more the machines we build and use resemble real persons in their behaviour, the harder it will become to escape that suggestive force. Yet while it is quite possible that we are easily fooled, that our natural constitution as human (or more generally animal) beings makes it rather difficult for us *not* to ascribe self-awareness to a machine that behaves exactly as we would expect it to if it were *really* self-aware, as long as we have an alternative explanation for why it behaves that way (namely that it has been designed and programmed to do so), we have no good reason to believe that its actions are expressive of anything at all. Even a perfect simulacrum is still a simulacrum, and our natural tendency to take the simulacrum for the real thing does nothing to change that.

Now, if James was right to surmise that what we want from a lover is that they *really* love us and not simply behave *as if* they loved us (while in fact not feeling anything at all), then a robot pseudo-person can never be as satisfactory as a human lover (at least not if we know that they are not human and have reason to believe or suspect that their apparent love is merely a clever simulation). Yet this also means that they can *only* be seen as satisfactory replacements for a human lover if all we care about, all we value, is what the other *does*, while not caring at all about how they *feel* or whether they feel anything at all. Human interaction is thus conceptually reduced to the behavioural aspect of it. True companionship where one person relates to another *through*, or by means of, their interactions is then no longer regarded as an end (because it is thought to be either unachievable or undesirable, or both). Instead, the means now *is* (understood to be) the end.

The attentive reader will have noticed that so far I have made no attempt to distinguish between love and sex, and I suspect that while many would agree that a robotic pseudo-person can never give us what we expect from somebody we *love* (namely that they love us back or, at the very least, that they are aware and appreciative of our love for them), the claim that such robots would be perfectly satisfactory as partners in a purely *sexual* relationship will be generally considered to be much more plausible. Yet the reason why I have avoided drawing a clear line between love and sex is not that I fail to acknowledge the difference. It is quite obvious to me that we can love someone without having sex with them, and have sex with them without loving them. So I am not conflating love with sex, nor do I think there is anything morally wrong with having casual sex, or sex without love. However, it seems to me that even what appears to be a purely sexual relationship between human partners is very often, and certainly can be, more than just sex (if we take "sex" to be a purely physical event). For one thing, it always, by necessity, involves an intimate encounter with another human being, a sharing of an experience. It is not merely the coming-together of two bodies that interact with each other. Rather, it is the interaction of two (or more) embodied *persons*. For another, when we have sex with another person, we are not, at least not normally, simply using the other person to, as it were, scratch a sexual itch. Sex is also about, and fuelled by, desire, and the knowledge or belief that this desire is reciprocated. We want to be or feel desired. We desire the other who desires us desiring them. Our lust and the pleasure we experience is at least partly a response to the lust and pleasure we incite in the other and to the lust and pleasure they desire to incite in us. Sex, or perhaps we should better say *good* sex—the kind that D.H. Lawrence used to call "tender-hearted fucking" (cf. Hauskeller 2014, 53–63)—is a practice of desire-sharing, a particular form of companionship and communion. In order to be fully satisfied with a robotic pseudo-person designed for sexual pleasure (who is, by definition, incapable of feeling any desire) we would have to attach no value to the interpersonal aspects of sex, that is, to those aspects of sex that can make it such a rich and exhilarating experience in the first place. This becomes quite evident when David Levy (2012, 227) declares that the "prime purpose of a sexbot is to assist the user in achieving orgasm, without the necessity of having another human being present". The human that is not present, and whose absence is supposedly fully compensated by the presence of the robot, is here seen as having the same function as the robot, namely to "assist the user in achieving orgasm". Not only does this view reduce the sexual act to what it often leads up to (as if nothing else mattered; the process itself discounted), it also assumes that to achieve full sexual satisfaction we do not need anybody else. All we need is someone or something (it doesn't matter which) that pushes the right buttons, that scratches what needs scratching and tickles what needs tickling. This someone could also be us. In other words, the other who is no longer a partner, but merely an "assistant", is nothing more than a rather overdeveloped masturbation device. You don't necessarily need a robot for that, and you certainly don't need a person. If sex is *in any case* nothing but masturbation (and at best mutual masturbation), then there is no reason to think that a pseudo-person, designed with sufficient technological sophistication, could not meet the job requirements just as well as a real person. But if sex is in fact more than that, or at least can be more than that, a communion of some sort, then sex with a pseudo-person can, just like masturbation, never be as satisfactory or fulfilling as sex with a real person.[1]

But what if we eventually managed to build robots that *are* real persons? Robots that can desire us as much as we can desire them, robots that can really love us back and feel what we feel. Would they then be just as good as a human lover? I am reasonably sure that for many, they would, provided they are, in all relevant respects, just like us (except perhaps better-looking and more skilled in the art of pleasuring the flesh). The fact that they would be human-made rather than human-born should not make a difference, although for some it might. Yet that would simply be a personal preference. Some might prefer synthetic lovers, others natural ones, just like some men prefer blondes and others brunettes. This does not say anything about their *general* preferability as sexual or romantic partners. However, that future social robots will in all relevant respects be like us is even more unlikely than that they will be persons, for the simple reason that they will in any case be made for a *purpose*, while humans generally are not (at least not yet). In order for them to exist, we will have to make them, and we are not going to do that without a good reason for it, and that means, without there being a need or demand for them. In other words, there has to be a market for them. So why would anyone want a robot lover? Why would anyone be willing to pay for them? Whitby (2012) lists several plausible motivations. Obviously, sexual companion robots might appeal to those who have trouble finding a human lover. Not everyone has the appearance or social skills that would make them attractive as a sexual partner to others, and even if they do find someone, those they can get may not be the ones they would have chosen if they had a choice. A sexual robot would allow those who are less sexually attractive not only to find a partner, but also to find a very attractive one. Others may simply like the idea of having sex with a machine (or in this case an artificial person). Possible reasons for this we have discussed in the fourth chapter of this book. Some people may feel drawn to the undemanding nature of robots that are designed to please us, and some may look forward to being able to do with their robotic partner whatever they want to without being restricted in any way by morality or by what their partner happens to like and dislike. But whatever the motivation, in order to give those people what they want, we can perhaps allow robots to be *persons* in the specified sense, but what we *cannot* allow is that they are free to act in a way that runs counter to the wishes of their buyers. If they think and feel, that's fine, perhaps even desirable, but they *must* love us when we want them to love us and have sex with us when we want them to have sex with us. Their freedom needs to be restricted. Otherwise, we would have no reason to create (and, perhaps more importantly, buy) them in the first place.

The required restriction of freedom can be achieved in two different ways. One option is to decide and decree that what the robot wants is of no significance and then to install a mechanism that prevents it from *doing* anything other than what we want it to do, either by programming it in such a way that it cannot disobey our commands (always assuming that this is possible, which it may well not be) or by creating a moral and legal framework that effectively leaves the robot no choice but to do our bidding (for fear of the repercussions that disobedience would incur). Both would amount to institutionally sanctioned slavery and might appeal to those who find rape (by which I here mean making someone have sex with you who does not want to have sex with you) more gratifying than consensual sex. The other, seemingly more morally palatable option is to design and programme robots in such a way that they never *want* to do anything other than what the buyer wants them to do. The first option has been suggested by Joanna Bryson (2010), and the second by Steve Petersen (2012).

According to Bryson, robots *should* be slaves. Not only would there be nothing morally wrong with keeping them as slaves, but also it would be morally wrong *not* to do so. It would be wrong to grant them any kind of moral status because doing so would draw time and energy, as well as care and emotional investment, away from those who deserve it, namely human persons. Ascriptions of personhood are a valuable resource with which we should not be too generous. And it would not be morally wrong to refuse robots moral consideration and keep them as slaves because we have created them specifically for that purpose, that is, to serve *our* needs and wants. For this reason, we should not have to treat them as persons or grant them any rights that we usually grant persons. As far as I can see, this claim is not based on the assumption that humanoid robots will not be real persons in the sense specified above, but only pseudo-persons. The term "person", for Bryson, seems to signify a being that deserves moral recognition. Personhood is here not a quality that an entity can possess, but something that is or is not *owed* to it. The term is thus purely normative and does not seem to have any descriptive dimension. Curiously, there is no suggestion in Bryson's paper that whether those robots are real persons or pseudo-persons in a *descriptive* sense might in any way be relevant for the question how we should regard and treat them. Rather, what settles the question for her is the fact that we have designed and produced them and therefore *own* them. There would be no robots without us, Bryson argues, they owe their existence solely to the fact that we needed and wanted somebody to perform a certain role. Since that is what they are here for, they are not entitled to demand or expect anything else from us. It does not really matter whether they are persons or pseudo-persons; what matters is that they are in either case still machines, made by us. And if we are, regardless, still afraid that an ethical issue might arise from enslaving them, we could just design them in such a way that they don't mind their lack of freedom. It is entirely up to us: "Remember, robots are wholly owned and designed by us. We determine their goals and desires. A robot cannot be frustrated unless it is given goals that cannot be met, and it cannot mind being frustrated unless we program it to perceive frustration as distressing, rather than as an indication of a planning puzzle." (Bryson 2010, 72). However, we, as their makers, have no direct moral obligation to them to spare them distress. At least that is how I understand Bryson's argument. If that is a fair reading, then I don't think the argument is very persuasive. The fact that an entity would not exist if we had not wanted it to exist, and to exist for a certain purpose, does not strip that entity of all moral standing. The same argument can, after all, also be made (and actually *has* been made) about the animals we breed for food and as pets. It may even be said about our own children. Yet since we don't usually accept this kind of reasoning (i.e. that what we create we own, and what we own has no rights), creating self-aware and self-concerned robot servants who do not want to serve us but have to do so anyway is not really an option.

To forestall such ethical concerns, Steve Petersen (2012) has suggested that we should design and programme our robot servants in such a way that they *want* to serve us, or more generally, that they want to do what we want them to do. So if nothing made them happier than to fulfil our every wish, then we would neither harm nor wrong them in any way by allowing them to do so. Nor would we wrong them by making it, right from the start, impossible for them to want anything else. If their wishes were always aligned with ours, then we would, Petersen argues, in fact not be treating them as mere means, because whatever we asked them to do would, per definition, benefit them just as much as us. It would not only serve *our* ends, but also *their* ends. Petersen does not think that we would thereby condemn them to a meaningless life. He rejects the idea that there are higher and lower pleasures, or pursuits that are more worth pursuing than others: instead, any pursuit must be regarded as worth pursuing as long as someone happens to get their kicks from it. Even our noblest pursuits are, after all, pretty meaningless in the grand scheme of things. And we do not usually think that the life of an animal (who is not capable of higher pursuits) is not worth living (although transhumanists, as we saw in the fifth chapter of this book, may well disagree). If we did, we should, or would, not allow them to exist in the first place.

While this is not the place to discuss in detail the ethics of creating artificial persons to serve us, a few remarks may be in order. If we accepted

Petersen's argument, then we would have to suppose that, for example, the life of a Sisyphus who was designed and programmed to find nothing more pleasurable than pushing a rock up a hill for his entire life is just as meaningful and fulfilling as the life of, say, a rescue worker who helps saving other people's lives. More importantly, we would also have to suppose that there is nothing wrong, nothing morally objectionable, with *deliberately creating* such a Sisyphus for our own ends (perhaps because we find it immensely amusing to see him pushing that bloody stone up the hill day after day, or, if we are more philosophically minded, to serve as a living reminder of the utter meaninglessness of all existence). Even though he may then not be doing anything that he does not want to do, we would still treat him merely as a means to our ends. To treat somebody as an end in itself does not merely mean that we let them do what they want to do, but also to allow them to want things that we do not want. It is about allowing someone to find their own ends without us making the decision for them. This is why it would also be wrong to breed or genetically engineer *human* persons who desire nothing more than to serve us, which, following Petersen's logic, would be morally unobjectionable, too.

Now imagine we had the means to create robotic or human persons whose only desire was to fulfil our sexual desires, whatever they may be. Custom-made models could be ordered online, fitted not only with specific bodily features, but also with particular preferential attitudes. People would get what they want, and what they want is someone who wants what they want them to want, for instance, "an airhead silicon bimboid obsessed with serving them sexually, or perhaps a skinnier anal-addicted Ukrainian model", or, for the more outlandish tastes, "babies for rape" or "snuff robots which scream and bleed realistically when their arms are sawn off" (Treanor 2015). All that would presumably be fine, ethically speaking (following Petersen) as long as those treated that way do not mind because it is what they themselves want anyway. Except it is *not* all right. It is demeaning, and the fact that we would have designed them to find pleasure in a demeaning life makes it not less but even more demeaning.[2]

Although the question I intended to address in this chapter is not whether it can be *morally acceptable* to have a sexual relationship with a robot, but whether such a relationship could ever be as *satisfying* as a sexual relationship with a human lover, the two questions are not as unconnected as it may seem. If that relationship does not agree with our ethical commitments, then we won't be able to regard and experience it as fully satisfying. That, of course, depends on our ethical commitments, whether we have any in the first place, and if yes, which. However, whether or not we have such commitments, the ethical features of a situation can still function as objective grounds for satisfaction and dissatisfaction in the sense that if a relationship is, to put it mildly, morally dubious, then, whether or not we are satisfied with it, it still remains the case that we *should* not be satisfied with it.

However, while most people will probably agree that the life of a person programmed to meet every and any sexual demand that we may have is indeed demeaning, this does not, or at least not so obviously, seem to be the case when a person is not designed for sex but for love. What if I just want somebody who loves me the way I am, someone who is good to me and there for me and will not tire of me and will not leave me because they find somebody who they think is more lovable? Surely there is nothing demeaning about loving a particular person and loving them reliably. So if it were possible to create such a person, and I ordered and purchased him or her, it does not seem that that person's life would be bad or meaningless because of it. (I am, after all, not such a bad guy and deserve to be loved by someone.) We may, of course, still take issue with the fact that they have been *designed* to love us and hence have no *choice* in the matter, but it is difficult to argue that point and I will not try to do this here.[3] So let us assume for now that creating persons programmed to love us is morally unobjectionable (as unobjectionable as, say, creating a person programmed with a burning desire to cure cancer). Would they then be as satisfying as a real (unprogrammed) human person who just happens to love us?

They might not, even if we have no ethical concerns about it. That is because, as Dylan Evans has pointed out, we do not only desire to be loved, but we usually also have the second-order desire to be loved freely, that is, by choice. "Although people typically want commitment and fidelity from their partners, they want these things to be the fruit of an ongoing choice, rather than inflexible and unreflexive behavior patterns" (Evans 2010, 81). We do not want people to love us because they have been hypnotised or enchanted (like Shakespeare's fairy queen Titania, who is made to fall in love with a donkey-headed weaver). And programming is, after all, just a different type of enchantment. Of course, we do not usually mind that those who love us "cannot help themselves", that their commitment to us is deeper and more unshakable than what a deliberate choice could provide (which can at any moment be revoked). Yet we do want the other to love us for what and who we are and not no matter what we are. We want it to be *their* choice and not ours (or, for that matter, a third party's). This entails a certain contingency, and with it, the possibility of loss. Even though we fear that possibility, we are unlikely to accord much value to a love that is ours whatever we do. We will probably tire of it very quickly. If that is correct, then an automatic sweetheart, even if they are real persons, will not be as satisfying as a human person who (really) loves us.

On the other hand, the prospect of having somebody that loves us *reliably*, someone who we *know* won't leave us no matter what, will certainly have a strong appeal to many. So there is indeed, as Evans puts it, a dilemma at the heart of the human–robot relationship: "We want contradictory things: a romantic partner who is both free and who will never leave us" (Evans 2010, 84). And if we cannot have both, then, depending on what we value most, we may well prefer the reliable artefact to the never-completely-reliable human. What is more, we might not even see this as a huge loss, or for that matter *any* loss, in the first place. We can, after all, always convince ourselves that nothing is really missing, that the robot gives us all that we can possibly get, or, at any rate, all that is worth having. The pseudo-person can be designed to appear and act like a real person, and if we cannot detect a difference and trust that "soul" is ultimately nothing but behaviour, then the pseudo-person will be just as good as a person. And if the robot *is* a real person, but not free to do anything but what we want them to do, then they can still *appear* free, and we can then, following the same reasoning, tell ourselves that they *are* free. It is easy to lie to ourselves if it gets us what we want: "[W]e are alone and imagine ourselves together" (Turkle 2011, 226).

Yet we may not even have to fool ourselves. Even if we are perfectly aware that the other that serves us as a pseudo-partner for sex and love does not really feel anything at all, or that if they do, they have been programmed to do so and hence have no real choice in the matter, we may actually *prefer* them that way. Being alone can be our preferred option. Unsurprisingly, it *is* the preferred option of the transhumanist. To engage with someone, a real human person, is, after all, always risky. Not only do we never quite know what we will get or whether we will actually get what it says on the tin, we are also constantly expected to take into account and sympathetically respond to *their* needs and desires. Real people are demanding and do not always perform the way we want them to. The great advantage of robots is that they do:

> Sexbots will never have headaches, fatigue, impotence, premature ejaculation, pubic lice, disinterest, menstrual blood, jock strap itch, yeast infections, genital warts, AIDS/HIV, herpes, silly expectations, or inhibiting phobias. Sexbots will never stalk us, rape us, diss us on their blog, weep when we dump them, or tell their friends we were boring in bed. Sexbots will always climax when we climax if we press that little button on their butt. (Pellissier 2009)

The author of this (by no means tongue-in-cheek) eulogy on sex robots, Hank Pellissier, used to be the managing director of the Institute for Ethics and Emerging Technologies (IEET), which has established itself as one of the two main transhumanist associations and think tanks (the other being Humanity Plus). The IEET's mission is to promote "ideas about how technological progress can increase freedom, happiness, and human flourishing".[4] Humanity Plus has a similar agenda: to elevate the human condition by expanding human capabilities and making us "better than well". Their motto is "Don't limit your challenges. Challenge your limits".[5] The suggestion that we can actually *benefit* from replacing human partners with robots must be understood in this context. Robots for sex and love constitute an important step towards the realisation of a shared transhumanist agenda, which, as we have seen, rests firmly on two ideological pillars: libertarianism and hedonism. From a transhumanist perspective, our goal should be to get the maximum amount of pleasure out of everything we do—which according to Nick Bostrom is, after all, nothing less than the "birth right of every creature" (Bostrom 2010, 6)—and to become as free/independent/autonomous as possible.

These two goals are connected, of course. Our limitations are thought to be a principal source of displeasure and unhappiness. Consequently, once we are free of all limitations, there will be nothing left to be unhappy about. Pellissier himself makes this connection explicit in a recent article, published on IEET's website. After examining the various kinds of "suffering caused by our enslavement to our outdated neurochemistry", he concludes that as "Free-will Transhumans, who decided 100 % of the time what we wanted to think, feel and do", we would not only be "immensely more powerful", but also, precisely for this reason, much happier (Pellissier 2015).

Now the problem with entering into relationships with other people is that, although they certainly can be a source of pleasure, more often than not they stand in the way of it. Moreover, even when they give us pleasure and happiness, this pleasure and happiness is always tainted and diminished by the fact that we need *them* to get it. Loving a human being and having sex with them might be pleasurable, but this pleasure can easily be taken away from us. From a transhumanist perspective, the fact that we *depend* on other people for sex and love is almost as annoying as the fact that we have to die or, more generally, the fact that we cannot do anything and everything that we want to do and not *be* anything and everything that we want to be. This (and not merely the fact that they might know better how to please and pleasure us) is the main reason why Levy thinks that sex and love robots are not only not deficient in any way, but actually *better* companions and lovers than a human could ever be.[6] Consider again the statement quoted earlier: "The prime purpose of a sexbot is to assist the user in achieving orgasm, without the necessity of having another human being present" (Levy 2012, 227). To have another human present is currently still a *necessity*, which is exactly what makes it problematic. Any necessity is bad because, by definition, it curbs our freedom. Necessities prevent us from being self-sufficient and truly autonomous. Sex robots are good not only because they are much more fun to be with, but also, even primarily, because they make us more independent. We should re-evaluate our attitude towards sex accordingly. "Are Sexbots icky? Are humans pathetic if we don't just mate with each other? Truth is, we're already mostly 'solo' when it comes to orgasms. 'Masturbation,' noted Hungarian psychiatrist Thomas Szasz, 'is the primary sexual activity…in the 19th century it was a disease, in the 20th it's a cure'" (Pellissier 2009).

A cure for what, though? Pellissier does not answer the question directly, but it is clear from the context that what this masturbation-by-sexbot is thought to be a cure for is the disease of other human beings (or the disease of our dependence on them). This gives a whole new meaning to Sartre's famous dictum that "hell is other people". The underlying logic is worrying. It hints at a paradox at the heart of the transhumanist agenda. If the goal is to increase my autonomy, and if other people, by virtue of having desires and needs of their own that differ from mine, necessarily impose limits on my autonomy, then in order to increase my own autonomy I need to find ways to *decrease* the autonomy of others, or, if that is not feasible, to create a world for myself that allows me to do what I want to without requiring the collaboration of others. As long as I have to interact with real others, as long as we share a world, our autonomy will always be severely restricted. Therefore, the only possible way for me to become completely independent is by cutting all ties to other persons, by making my own world, uninhabited by any real persons except myself. Perfect autonomy (and thus supposedly perfect happiness) requires complete detachment. Robotic pseudo-persons or persons can then be understood as an enhanced version of other people. They are in fact, in more than one sense, postpersons.

Notes

1. George Mikes, in his humorous 1946 classic *How to Be an Alien*, quipped that "continental people have a sex life; the English have hot-water bottles" (Mikes 1986, 25). Sex robots are something like the hot-water bottles of the future.
2. John Danaher has recently (2016) discussed how it may be considered morally wrong (and hence apt for criminalisation) to engage in sexual activities with robots deliberately designed to cater to "paedophilic tastes and rape fantasies". Danaher assumes that those robots will be pseudo-persons, which makes it more difficult to establish any wrongdoing. However, he argues (although rather tentatively) that such activities may reasonably be regarded

as harmful to the moral character of those who engage in them. Although I think more needs to be said about what that means exactly, the point is well-taken. Perhaps the supposed harm to moral character can be best understood in relation to the self-demeaning nature of the activity. And if the robot subjected to these activities is a person, then what is happening is demeaning for both.

3. I have tried to make that argument in Hauskeller (2016a).
4. http://www.ieet.org/index.php/IEET/about
5. http://humanityplus.org/about/
6. Evans (2010) fittingly calls this the "greater satisfaction thesis".

Bibliography

Bostrom, Nick (2010) "A Letter from Utopia", version 1.9: http://www.nick-bostrom.com/utopia.pdf.

Bryson, Joanna J. (2010) "Robots Should Be Slaves", in *Close Engagements with Artificial Companions. Key social, psychological, ethical and design issues*, ed. Yorick Wilks, Amsterdam/Philadelphia: John Benjamins Publishing Company, 63–74.

Danaher, John (2016) "Robotic Rape and Robotic Child Sex Abuse: Should they by criminalised?", *Criminal Law and Philosophy* (DOI: 10.1007/s11572-014-9362-x, forthcoming, 2016).

Evans, Dylan (2010) "Wanting the Impossible. The dilemma at the heart of intimate human-robot relationships", in *Close Engagements with Artificial Companions. Key social, psychological, ethical and design issues*, ed. Yorick Wilks, Amsterdam/Philadelphia: John Benjamins Publishing Company, 75–74.

Hauskeller, Michael (2014) *Sex and the Posthuman Condition*, Basingstoke: Palgrave Macmillan.

——— (2016a): "Is It Desirable to Be Able to Do the Undesirable", *Cambridge Quarterly of Healthcare Ethics* (forthcoming).

James, William (1909) *The Meaning of Truth. A Sequel to 'Pragmatism'*, New York, Bombay and Calcutta: Longmans, Green and Co.

Kurzweil, Ray (1999) *The Age of Spiritual Machines*, New York: Penguin.

Levy, David (2008) *Love & Sex with Robots. The Evolution of Human-Robot Relationships*, London: Duckworth Overlook.

——— (2012) "The Ethics of Robot Prostitutes", in *Robot Ethics. The Ethical and Social Implications of Robotics*, eds. Patrick Lin, Keith Abney, and George A. Bekey, Cambridge, MA: MIT Press, 223–31.

Mikes, George (1986) *How to Be a Brit*, Harmondsworth: Penguin.

Pellissier, Hank (2009) "Sexbots will give us longevity orgasm", *Humanity plus Magazine*, 11 December 2009, http://hplusmagazine.com/2009/12/11/sexbots-will-give-us-longevity-orgasm/

——— (2015) Free Will Does Not Exist – Should it be a Transhumanist Enhancement?", http://ieet.org/index.php/IEET/more/pellissier20150727

Petersen, Steve (2012) "Designing People to Serve", in Robot Ethics. The Ethical and Social Implications of Robotics, eds. Patrick Lin, Keith Abney, and George A. Bekey, Cambridge, MA: MIT Press.

Pew Research Center (2014): "AI, Robotics, and the Future of Jobs", http://www.pewinternet.org/2014/08/06/future-of-jobs/

Scheutz, Matthias (2012) "The Inherent Dangers of Unidirectional Emotional Bonds between Humans and Social Robots", in *Robot Ethics. The Ethical and Social Implications of Robotics*, eds. Patrick Lin, Keith Abney, and George A. Bekey, Cambridge, MA: MIT Press.

Singer, Edgar Arthur (1924) *Mind as Behavior and Studies in Empirical Idealism*, Columbus, OH: R.G. Adams and Co.

Treanor, Paul (2015) "Bimboid utopia: what do men want?", https://politicalaspects.wordpress.com/2015/02/06/bimboid-utopia-what-do-men-want/

Turkle, Sherry (2011) *Alone Together. Why We Expect More from Technology and Less from Each Other*, New York: Basic Books.

Whitby, Blay (2012) "Do You Want a Robot Lover? The Ethics of Caring Technologies", in *Robot Ethics. The Ethical and Social Implications of Robotics*, eds. Patrick Lin, Keith Abney, and George A. Bekey, Cambridge, MA: MIT Press, 234–248.

Epilogue: Flowers from the Future

I must have been seven or eight years old when I first met the time traveller. My parents and I were spending the day at my grandparents', as we did on most Sundays, and as usual, I was bored stiff because nobody would talk to me, my grandparents didn't have any games or toys, and there were no books that invited me to read them. All they had was a mean old dog who barked madly and threatened to bite me whenever I tried to get up from the sofa or made a movement that alerted him to the fact that I was still alive. He seemed determined to change that. But on this particular day, I was lucky. Not only was the dog exhausted from a long walk in the park and for the time being showed no interest in harassing me any further, but I was also granted permission to switch on the television. They were showing a film, and it had just started. It was *The Time Machine*, the 1960 version with Rod Taylor, and for the next 90 minutes or so, I was lost to the world. I don't think I have ever seen a film that impressed me as much as this one. I was not only fascinated by the idea of time travelling, but also utterly enthralled by the story woven around it and the images that brought the story to life. For many weeks after, I had recurring, nightmarish dreams of the beautiful and gentle, but

M. Hauskeller, *Mythologies of Transhumanism*,
DOI 10.1007/978-3-319-39741-2

rather dumb Eloi and the creepy, demonic Morlocks with their greenish skin and red eyes, who scared the hell out of me.

It was many years later that I finally read the book by H.G. Wells on which the film was based. I was already in my twenties, and by then, the idea of time travel had lost some of its early fascination for me. The Morlocks, too, were rather disappointing. So I cannot really say that, at that time, the book made a lasting impression on me. However, when I read it again a couple of years ago, I was surprised to see what a marvellous writer Wells actually was. And the future of humanity that he shows us is a far cry from the superglossed one that transhumanists and other enhancement enthusiasts keep dangling in front of our noses. What awaits us in his vision is not an "engineered Paradise" (David Pearce), nor "lives wonderful beyond imagination" (Nick Bostrom), even though at first glance it looks like a paradise and it was actually meant to be one. It is a dystopia that started out as a eutopia, and that still disguises itself as one. But Wells shows us that every paradise has a dark side, that there is always a price to pay, and that what seems to be progress may well prove to be humanity's downfall. It is quite likely that as a species, we will develop further, and it is even possible that we will be able to steer that development in the direction that appears most desirable to us. But that doesn't mean that we will like what we will get. We tend to think of the posthuman as something that is better than a mere human, more advanced, an improved human. But the posthuman may just as well turn out to be in some important respect less than human.

But of course, for Wells, it is even worse than that. Wells's time traveller travels further and further into the future until eventually even those shrunken versions of our present selves have vanished. Wells allows us a glimpse of a time when we will all be gone for good and there will be nothing left. It will be as if we had never existed, the world an empty, desolate place. No humans, no post-humans, nothing, just tohu wa-bohu. It is a truly chilling prospect, masterly set on scene by Wells:

> The darkness grew apace; a cold wind began to blow in freshening gusts from the east, and the showering white flakes in the air increased in number. From the edge of the sea came a ripple and whisper. Beyond these lifeless sounds the world was silent. Silent? It would be hard to convey the

> stillness of it. All the sounds of man, the bleating of sheep, the cries of birds, the hum of insects, the stir that makes the background of our lives – all that was over. As the darkness thickened, the eddying flakes grew more abundant, dancing before my eyes; and the cold of the air more intense. At last, one by one, swiftly, one after the other, the white peaks of the distant hills vanished into blackness. The breeze rose to a moaning wind. I saw the black central shadow of the eclipse sweeping towards me. In another moment the pale stars alone were visible. All else was rayless obscurity. The sky was absolutely black.

Surprisingly, however, the novel ends on an optimistic note. The time traveller has brought home from his journey two flowers from the pre-desolation future and passes them on to the story's narrator before he leaves once more, never to return again. Those two flowers provide some comfort to the narrator, despite everything that is going to happen, because they remind him of what truly matters in life: "And I have by me, for my comfort, two strange white flowers – shrivelled now, and brown and flat and brittle – to witness that even when mind and strength had gone, gratitude and a mutual tenderness still lived on in the heart of man."

In the longest run, our prospects may be very bleak indeed, but as long as we can hold on to that "mutual tenderness", all is not lost.

Bibliography

Agar, Nicholas (2004) *Liberal Eugenics*, London: Blackwell.

Agar, Nicholas (2010) *Humanity's End*, Cambridge, MA: MIT Press.

Agar, Nicholas (2013) "Why it is possible to enhance moral status and why doing so is wrong", *Journal of Medical Ethics* 39: 67–74.

Agar, Nicholas (2014) *Truly Human Enhancement: A Philosophical Defense of Limits*, Cambridge, MA: MIT Press.

Anders, Günther (1956) *Die Antiquiertheit des Menschen*, Munich: C.H. Beck.

Ansell-Pearson, Keith (1992) "Who is the Ubermensch? Time, Truth, and Woman in Nietzsche", *Journal of the History of Ideas* 53/2: 309–331.

Atwood, Margaret (2009) *The Year of the Flood*, London: Bloomsbury.

Aquinas, Thomas (1980), ed. Charles H. Lohr, *Scriptum super Sententiis*, Fordham University Press.

Bacon, Francis (1924) *New Atlantis*, Oxford: Clarendon Press.

Banja, John (2011) "Virtue essentialism, prototypes, and the moral conservative opposition to enhancement: A neuroethical critique", *AJOB Neuroscience* 2/2: 31–38.

Bashford, Alison (2013) "Julian Huxley's Transhumanism", in *Crafting Humans. From Genesis to Eugenics and Beyond*, ed. Marius Turda, National Taiwan University Press, 153–167.

Benatar, David (2006) *Better Never to Have Been. The Harm of Coming into Existence*, Oxford: Oxford University Press.

M. Hauskeller, *Mythologies of Transhumanism*,
DOI 10.1007/978-3-319-39741-2

Benatar, David, and David Wasserman (2015) *Debating Procreation. Is It Wrong to Reproduce*, Oxford: Oxford University Press.

Berthoud, Gerald (2007) "The Techno-Utopia of 'Human Performance Enhancement", in: *Utopie heute 1*, ed. Beat Sitter-Liver, Fribourg: Academic Press/Stuttgart: Kohlhammer, 291–305.

Borges, Jorge Luis (1962) *Labyrinths*, New York: New Directions Publishing.

Bostrom, Nick (2004) "Golden", www.nickbostrom.com.

Bostrom, Nick (2005) "The Fable of the Dragon Tyrant," *Journal of Medical Ethics* 31: 273–277.

Bostrom, Nick (2005a) "A History of Transhumanist Thought", *Journal of Evolution and Technology* 14/1: 1–25.

Bostrom, Nick (2005b) "Transhumanist Values", *Review of Contemporary Philosophy* 4.

Bostrom, Nick (2009) "Why I Want to be a Posthuman When I Grow Up", in *Medical Enhancement and Posthumanity*, eds. B. Gordijn and R. Chadwick, Dordrecht: Springer, 107–137.

Bostrom, Nick (2010) "A Letter from Utopia", version 1.9: http://www.nickbostrom.com/utopia.pdf.

Bostrom, Nick (2014) *Superintelligence*, Oxford: Oxford University Press.

Boswell, James (1924) *Journal of a Tour to the Hebrides with Samuel Johnson, LL.D.*, ed. R.W. Chapman, London.

Bradley, Ben (2004) "When Is Death Bad for the One Who Dies?", *Nous* 38/1: 1–28.

Brandow, Michael (2015) *A Matter of Breeding*, Boston: Beacon Press.

Brin, David (2012) *Uplift. The Complete Original Trilogy*, London: Orbit.

Bryson, Joanna J. (2010) "Robots Should Be Slaves", in *Close Engagements with Artificial Companions. Key social, psychological, ethical and design issues*, ed. Yorick Wilks, Amsterdam/Philadelphia: John Benjamins Publishing Company, 63–74.

Buchanan, Allen (1995), "Equal Opportunity and Genetic Intervention", *Social Philosophy and Policy Foundation* 12/2: 105–135.

Buchanan, Allen, Dan W. Brocks, Norman Daniels, and Daniel Wikler (2000) *From Chance to Choice. Genetics & Justice*, Cambridge University Press.

Buchanan, Allen (2011) *Beyond Humanity?* Oxford: Oxford University Press.

Burdett, Michael S. (2011) "Contextualizing a Christian Perspective on Transcendence and Human Enhancement", in *Transhumanism and Transcendence*, ed. Ronald Cole-Turner, Washington, DC: Georgetown University Press, 19–35.

Burdett, Michael S. (2015) "The Religion of Technology: Transhumanism and the Myth of Progress", in *Religion and Transhumanism. The Unknown Future of Human Enhancement*, eds. Calvin Mercer and Tracy J. Trothen, Santa Barbara: Praeger, 131–147.

Carel, Havi (2013) *Illness. The Cry of the Flesh*, Revised edition, Durham: Acumen.

Carlyle, Thomas (1908) *Sartor Resartus*, London: Everyman's Library.

Carlyle, Thomas (1970) *The Collected Letters of Thomas and Jane Welsh Carlyle*, vol. 1, Durham, NC.

Chan, S (2009) "Should we enhance animals?", *Journal of Medical Ethics* 35: 678–683.

Chesterton, Gilbert Keith (1908) *Orthodoxy*, London: Bodley Head.

Condorcet, Marquis de (1955) *Sketch for a Historical Picture of the Progress of the Human Mind*, London: Weidenfeld & Nicolson.

Coudert, Allison (1980) *Alchemy: the Philosopher' Stone*, London: Wildwood House.

Crevier, Daniel (1993) *AI: The Tumultuous Search for Artificial Intelligence*. New York: Basic Books.

Cuthbertson, Anthony (2015) "Virtual reality heaven: How technology is redefining death and the afterlife", International Business Times, 9 December 2015: http://www.ibtimes.co.uk/virtual-reality-heaven-how-technology-redefining-afterlife-1532429

Danaher, John (2016) "Robotic Rape and Robotic Child Sex Abuse: Should they by criminalised?", *Criminal Law and Philosophy* (DOI: 10.1007/s11572-014-9362-x, forthcoming, 2016).

Deane-Drummond, Celia (2011) "Taking Leave of the Animal? The Theological and Ethical Implications of Transhuman Projects", in *Transhumanism and Transcendence*, ed. Ronald Cole-Turner, Washington, DC: Georgetown University Press, 115–130.

DeBaets, Amy Michelle (2015) "Rapture of the Geeks: Singularitarianism, Feminism, and the Yearning for Transcendence", in *Religion and Transhumanism. The Unknown Future of Human Enhancement*, eds. Calvin Mercer and Tracy J. Trothen, Santa Barbara: Praeger, 181–197.

De Grey, Aubrey, and Michael Rae (2007) *Ending Aging*, New York: Saint Martin's Press.

Derry, Margaret E. (2003) *Bred for Perfection. Shorthorn Cattle, Collies and Arabian Horses since 1800*, Baltimore/London: John Hopkins University Press.

Diogenes Laertius (1958) *Lives of Eminent Philosophers*, London: William Heinemann/Cambridge, MA: Harvard University Press.

Dupre, John (2001) *Human Nature and the Limits of Science*, Oxford: Clarendon Press.

Duncan, Collin (2015) "The Day no one died: The Moral Imperative of Transhumanism (https://medium.com/@BleachedSleet/the-day-no-one-died-the-moral-imperative-of-transhumanism-5d74d6c3c234#.mi709bp9n)

Dvorsky, George (2008) "All Together Now: Developmental and ethical considerations for biologically uplifting nonhuman animals", *Journal of Evolution and Technology* 18/1: 129–140.

Dvorsky, George (2012) "Should we upgrade the intelligence of animals?", *io9 blog* 9.7.2012, http://io9.gizmodo.com/5943832/should-we-upgrade-the-intelligence-of-animals.

Earp, Brian D., Olga A. Wudarczyk, Anders Sandberg, and Julian Savulescu (2013) "If I Could Just Stop Loving You: Anti-Love Biotechnology and the Ethics of a Chemical Breakup", *The American Journal of Bioethics* 13/11: 3–17.

Earp, Brian D., Anders Sandberg, and Julian Savulescu (2015) "The Medicalization of Love", *Cambridge Quarterly of Healthcare Ethics* 24/3: 323–336.

Earp, Brian D., Olga A. Wudarczyk, Bennett Foddy, and Julian Savulescu (2016), "Addicted to Love: What Is Love Addiction and When Should It Be Treated?", *Philosophy, Psychiatry, & Psychology*, in press.

Eurich, Nell (1967) *Science in Utopia*, Cambridge, MA: Harvard University Press.

Evans, Dylan (2010) "Wanting the Impossible. The dilemma at the heart of intimate human-robot relationships", in *Close Engagements with Artificial Companions. Key social, psychological, ethical and design issues*, ed. Yorick Wilks, Amsterdam/Philadelphia: John Benjamins Publishing Company, 75–74.

Ferguson, Anthony (2010) *The Sex Doll. A History*, Jefferson & London: McFarland & Co.

Ferrari, Arianna (2012) "Animal Disenhancement for Animal Welfare: The Apparent Philosophical Conundrum and the Real Exploitation of Animals", *Nanoethics* 6: 65–76.

Ferrari, Arianna (2015) "Animal Enhancement: Technovisionary Paternalism and the Colonisation of Nature", in *Inquiring into Animal Enhancement: Model our Countermodel of Human Enhancement?*, eds. Simone Bateman et al, Basingstoke: Palgrave Macmillan, 13–33.

Fichte, Johann Gottlieb (1908) *The Vocation of Man*, London: John Chapman.

Frommherz, Gudrun (2013) "Memetics of Transhumanist Imagery", *Visual Anthropology* 26/2: 147–164.

Fudge, Erica (2006) *Brutal Reasoning. Animals, Rationality, and Humanity in Early Modern England*, Ithaca and London: Cornell University Press.

Fukuyama, Francis (2002) *Our Posthuman Future*, New York: Farrar, Straus & Giroux.

Fukuyama, Francis (2009) Transhumanism, *Foreign Policy*, http://foreignpolicy.com/2009/10/23/transhumanism/

Fuller, Steve, and Veronika Lipinska (2014) *The Proactionary Imperative. A Foundation for Transhumanism*, Basingstoke: Palgrave Macmillan.

Frye, Northrop (1957) *Anatomy of Criticism*, Princeton, NJ: Princeton University Press.

Gane, Nicholas/Haraway, Donna (2006) "When we have never been human, what is to be done? Interview with Donna Haraway", *Theory Culture Society* 23, 135–158.

Gehlen, Arnold (1940) *Der Mensch. Seine Natur und seine Stellung in der Welt*, Berlin: Junker und Dünnhaupt.

Gilbert, Paul (2015) "'You love life, but we love death': Suicide Bombings and Drone Attacks" in *Making Humans*, ed Alexander Darius Ornella, Inter-Disciplinary Press.

Gillman, Neil (2004) "A Jewish Theology of Death and the Afterlife", in: *The Fountain of Youth*, ed. Stephen G. Post and Robert H. Binstock, Oxford: Oxford University Press, 94–108.

Graham, Elaine L. (2002) *Representations of the Post/Human*, New Brunswick: Rutgers University Press,

Godwin, William (1971) *An Enquiry Concerning Political Justice*, London: Oxford University Press.

Greenemeier, Larry (2011) "Robotic Exoskeletons from Cyberdyne Could Help Workers Clean Up Fukushima Nuclear Mess." *Scientific American* 9/11/11 http://blogs.scientificamerican.com/observations/2011/11/09/robotic-exoskeletons-from-cyberdyne-could-help-workers-clean-up-fukushima-nuclear-mess/.

Gupta, Sanjay (2012) Interview with Hugo Herr. *CNN Transcripts* 25/3/12 http://transcripts.cnn.com/TRANSCRIPTS/1203/25/nl.01.html.

Habermas, Jürgen (2003) *The Future of Human Nature*, Cambridge: Polity Press.

Halberstam, Judith, and Ira Livingston (eds.) (1995) *Posthuman Bodies*, Bloomington and Indianapolis: Indiana University Press.

Haraway, Donna (1985) "A Manifesto for Cyborgs. Science, Technology, and Socialist Feminism in the 1980s", in: *Feminism/Postmodernism*, ed. L. Nicholson: New York and London: Routledge 1990, 190–233.

Haraway, Donna (2008) *When Species Meet*, Minneapolis/London: University of Minnesota Press.

Harrington, Alan (1969) *The Immortalist*, New York: Random House.

Harris, John (1993) "Is Gene Therapy a Form of Eugenics", *Bioethics* 7/2-3: 178–87.

Harris, John (2007) *Enhancing Evolution. The Ethical Case for Making Better People*, Princeton: Princeton University Press.

Harris, Steven B. (1996) "The Immortality Myth and Technology", in: *Immortal Engines. Life Extension and Immortality in Science Fiction and Fantasy*, ed. G. Slusser, G. Westfahl Westfahl, and E.S. Rabkin, Athens and London: University of Georgia Press 1996, 45–67.

Hauskeller, Michael (2007) "A Legacy of Love? Reflections on the New "Humanitarian" Eugenics," in *Utopie heute 1*, ed B. Sitter-Liver, Fribourg: Academic Press/Stuttgart: Kohlhammer, 405–419.

Hauskeller, Michael (2007a) *Biotechnology and the Integrity of Life*, Aldershot: Ashgate.

Hauskeller, Michael (2009) "Prometheus Unbound. Transhumanist Arguments from (Human) Nature", *Ethical Perspectives* 16/1: 3–20.

Hauskeller, Michael (2011) "Is Ageing Bad for Us?" *Ethics and Medicine* 27/1: 25–32.

Hauskeller, Michael (2011a) "Forever Young? Life Extension and the Ageing Mind", *Ethical Perspectives* 18/3: 385–406.

Hauskeller, Michael (2011b) "Human Enhancement and the Giftedness of Life", *Philosophical Papers* 40/1: 55–79.

Hauskeller, Michael (2012) "My Brain, my Mind, and I. Some Philosophical Assumptions of Mind-Uploading." *International Journal of Machine Consciousness* 4/1: 187–200.

Hauskeller, Michael (2013) *Better Humans? Understanding the Enhancement Project*, London: Routledge.

Hauskeller, Michael (2014) *Sex and the Posthuman Condition*, Basingstoke: Palgrave Macmillan.

Hauskeller, Michael (2016) "Topsyturvy – Jonathan Swift on Human Nature, Reason, and Morality", in Janelle Poetzsch, *Jonathan Swift and Philosophy*.

Hauskeller, Michael (2016a): "Is It Desirable to Be Able to Do the Undesirable", *Cambridge Quarterly of Healthcare Ethics* (forthcoming).

Hayles, N. Katharine (1999) *How We Became Posthuman*, Chicago and London: The University of Chicago Press.

Heimerl, Theresa (2015) "Ascetic Athletes: The Hero-Body in Contemporary Hollywood Cinema", in *Making Humans*, ed. Alexander Darius Ornella, Interdisciplinary Press.

Henschke, Adam (2012) "Making Sense of Animal Disenhancement", *Nanoethics* 6: 55–64.

Hesiod (1993) *Works and Days*, Indianapolis: Hackett Publishing.

Holtug, Nils (1999) "Does Justice Require Genetic Enhancements?", *Journal of Medical Ethics* 25: 137–143.

Hooker, Brad (2005) "Fairness", *Ethical Theory and Moral Practice* 8: 329–352.

Hopkins, Patrick D. (2005) "Transcending the Animal. How Transhumanism and Religion Are and Are Not Alike", *Journal of Evolution and Technology* 14/2: 13–28.

Hughes, James (2004) *Citizen Cyborg. Why Democratic Societies Must Respond to the Redesigned Human of the Future*, Cambridge, MA: Westview Press.

Hughes, James (2013) "Transhumanism and Personal Identity", in *The Transhumanist Reader*, eds. Max More and Natasha Vita-More, Chichester: Wiley Blackwell, 227–233.

Hull, David (1998) "On Human Nature", in: *The Philosophy of Biology*, eds. David Hull and Michael Rose, Oxford: Oxford University Press, 383–397.

Humanity Plus (1998) *Transhumanist Declaration*, http://humanityplus.org/philosophy/transhumanist-declaration/

Huxley, Julian (1957) "Transhumanism", in Julian Huxley, *New Bottles for New Wine*, London: Chatto & Windus, 13–17.

Istvan, Zoltan (2013) *The Transhumanist Wager*, Futurity Imagine Media.

James, William (1909) *The Meaning of Truth. A Sequel to 'Pragmatism'*, New York, Bombay and Calcutta: Longmans, Green and Co.

Jaspers, Karl (1971) *Philosophy of Existence*, Philadelphia.

Jonas, Hans (1966) *The Phenomenon of Life*, New York.

Jonas, Hans (1985) "Werkzeug, Bild und Grab", *Scheidewege* 15: 85–112.

Kafka, Franz (1915) *Metamorphosis*, https://www.gutenberg.org/files/5200/5200-h/5200-h.htm

Kafka, Franz (1917) *A Report for an Academy*, http://records.viu.ca/~johnstoi/kafka/reportforacademy.htm

Kass, Leon (2002) "L'Chaim and its Limits: Why not Immortality?", in: Leon Kass, *Life, Liberty and the Defence of Dignity: The Challenge for Bioethics*, San Francisco: Encounter, 257–274.

Kass, Leon (2003) "Ageless bodies, happy souls", *New Atlantis* 1/1: 9–28.

Kitcher, Philip (1985) *Vaulting Ambition*, Cambridge, MA, London: MIT Press.

Kurzweil, Ray (1999) *The Age of Spiritual Machines*, New York: Penguin.

Kurzweil, Ray (2005) *The Singularity is Near*, New York: Viking Adult.

Lampert, Laurence (1987) *Nietzsche's Teaching. An Interpretation of Thus Spoke Zarathustra*, New Haven: Yale University Press.

Levy, David (2008) *Love & Sex with Robots. The Evolution of Human-Robot Relationships*, London: Duckworth Overlook.
Levy, David (2012) "The Ethics of Robot Prostitutes", in *Robot Ethics. The Ethical and Social Implications of Robotics*, eds. Patrick Lin, Keith Abney, and George A. Bekey, Cambridge, MA: MIT Press, 223–31.
Lewens, Tim (2009) "What Are 'Natural Inequalities'?" *The Philosophical Quarterly* 60/239: 264–285.
Lewis, C.S. (1955) *The Abolition of Man* [1943], New York: Macmillan.
Locke, John (1823) *An Essay Concerning Human Understanding*, in The Works of John Locke. A New Edition, Corrected. In Ten Volumes, vol. II, London 1823.
Marks, Jonathan (2009) "The Nature of Humanness", in *Oxford Handbook of Archaelogy*, eds. Chris Gosden, Barry Cunliffe and Rosemary A. Joyce, Oxford: Oxford University Press, 237–53.
McKibben, Bill (2003) *Enough. Genetic engineering and the end of human nature*, London: Bloomsbury.
Miah, Andy (2008) "A Critical History of Posthumanism", in *Medical Enhancement and Posthumanity*, eds. Bert Gordijn/Ruth Chadwick, New York: Springer, 71–94.
Midgley, Mary (2011) *Myths We Live By*, with a new foreword by the author, London: Routledge Classics.
Mikes, George (1986) *How to Be a Brit*, Harmondsworth: Penguin.
Mizrahi, Moti (2014) "The Problem of Natural Inequality: A New Problem of Evil", *Philosophia* 42: 127–136.
Moravec, Hans (1988) *Mind Children. The Future of Robot and Human Intelligence*, Cambridge, MA: Harvard University Press.
More, Max (1995) *The Diachronic Self: Identity, Continuity, Transformation*, Ann Arbor, MI: A Bell and Howell Company.
More, Max (1996) "Transhumanism. Towards a Futurist Philosophy", www.maxmore.com/transhum.htm.
More, Max (2009) "True Transhumanism", *Global Spiral*, February.
More, Max (2010) "The Overhuman in the Transhuman", *Journal of Evolution and Technology* 21/1: 1–4.
More, Max (2013) "The Proactionary Principle. Optimizing Technological Outcomes", in: *The Transhumanist Reader*, eds. Max More and Natasha Vita-More, Chichester: Wiley-Blackwell, 258–267.
More, Max (2013a) "The Philosophy of Transhumanism", in: *The Transhumanist Reader*, eds. Max More and Natasha Vita-More, Chichester: Wiley-Blackwell, 3–17.

More, Max (2013b) "A Letter to Mother Nature", in: *The Transhumanist Reader*, eds. Max More and Natasha Vita-More, Chichester: Wiley-Blackwell, 449–450.

More, Max, and Natasha Vita-More (eds) (2013), *The Transhumanist Reader*, Chichester: Wiley Blackwell

Morgan, Kathryn P. (1998) "Women and the Knife: Cosmetic Surgery and the Colonization of Women's Bodies", in: *Sex/Machine: Readings in Culture, Gender, and Technology*, ed. P.D. Hopkins, Bloomington: Indiana University Press, 261–285.

Muller, Christopher John, and Günther Anders (2016) *Promethean Technology, Digital Culture and Human Obsolescence*, London: Rowman & Littlefield International.

Nagel, Thomas (1979) "Death", in Thomas Nagel, *Mortal Questions*, New York: Cambridge University Press, 1–10.

Nelkin, Dorothy, and Susan Lindee (1995) *The DNA Mystique: The Gene as a Cultural Icon*, New York: Freeman.

Nietzsche, Friedrich (1966) *Werke in drei Bänden* (W). Ed. Karl Schlechta, Munich: Hanser Verlag.

Noonan, John T. junior (1993) "An Almost Absolute Value in History", in: *Arguing About Abortion*, ed. Lewis M. Shwartz, Belmont, CA: Wadsworth, 51–59.

Nussbaum, Martha (1994) *The Therapy of Desire: Theory and Practice in Hellenistic Ethics*, Updated Edition 2009, Princeton: Princeton University Press.

Onstott, Kyle (1975) *The New Art of Breeding Better Dogs*, New York: Howell Book House.

Palmer, Clare (2011) "Animal Disenhancement and the Non-Identity Problem: A Response to Thompson, *Nanoethics* 5: 43–48.

Parens, Erik (1998) "Is Better Always Good? The Enhancement Project", in *Enhancing Human Traits: Ethical and Social Implications*, ed Erik Parens, Washington D.C.: Georgetown University Press.

Parker, Lisa (2012) "In Sport and Social Justice, Is Genetic Enhancement a Game Changer?", *Health Care Analysis* 20: 328–346.

Paul, Gregory S., and Earl Cox (1996) *Beyond Humanity*, Rockland, MA: Charles River.

Pearce, David (1995) *The Hedonistic Imperative*, www.hedweb.com.

Pearce, David (2007) "The Abolitionist Project", www.abolitionist.com.

Pellissier, Hank (2009) "Sexbots will give us longevity orgasm", *Humanity plus Magazine*, 11 December 2009, http://hplusmagazine.com/2009/12/11/sexbots-will-give-us-longevity-orgasm/

Pellissier, Hank (2015) Free Will Does Not Exist – Should it be a Transhumanist Enhancement?", http://ieet.org/index.php/IEET/more/pellissier20150727

Persson, Ingmar, and Julian Savulescu (2008) "The Perils of Cognitive Enhancement and the Urgent Imperative to Enhance the Moral Character of Humanity", *Journal of Applied Philosophy* 25/3: 162–77.

Persson, Ingmar, and Julian Savulescu (2012) *Unfit for the Future: The Need for Moral Enhancement*, Oxford: Oxford University Press.

Petersen, Steve (2012) "Designing People to Serve", in *Robot Ethics. The Ethical and Social Implications of Robotics*, eds. Patrick Lin, Keith Abney, and George A. Bekey, Cambridge, MA: MIT Press, 283–298.

Pew Research Center (2014): "AI, Robotics, and the Future of Jobs", http://www.pewinternet.org/2014/08/06/future-of-jobs/

Pico della Mirandola (1985) *On the Dignity of Man, On Being and the One, Heptaplus*, New York: Macmillan.

Pleij, Herman (2001) *Dreaming of Cockaigne*, New York: Columbia University Press.

Prisco, Giulio (2015) "The Quest for Akashic Physics", *Turing Church*, 28 November 2015: http://turingchurch.com/2015/11/28/the-quest-for-akashic-physics/#more-3603

Rawls, John (1999) *A Theory of Justice. Revised Edition*, Cambridge, MA: Belknap Press.

Riskin, Jessica (2003a) "Eighteenth-Century Wetware", *Representations* 83: 97–125.

Riskin, Jessica (2003b): "The Defecating Duck, or, the Ambiguous Origins of Artificial Life", *Critical Inquiry* 29/4: 599–633.

Robert, Jason Scott, and Francoise Baylis (2003), "Crossing Species Boundaries", *The American Journal of Bioethics* 3/3: 1–13

Roco, Mihail C., and William Sims Bainbridge (eds.) (2003) *Converging Technologies for Improving Human Performance*, Dordrecht: Springer.

Rollin, Bernard (1995) *The Frankenstein Syndrome. Ethical and Social Issues in the Genetic Engineering of Animals*, Cambridge: Cambridge University Press.

Rorabaugh W.J. (2004) *Kennedy and the Promise of the Sixties*, Cambridge University Press.

Rosenfield, Leonora Cohen (1968) *From Beast-Machine to Man–machine. Animal Soul in French Letters from Descartes to La Mettrie*, New York: Octagon Books.

Sandberg, Anders, and Nick Bostrom (2008) *Whole Brain Emulation. A Road Map*. Oxford: Future of Humanity Institute, Oxford University.

Sandberg, Anders (2012) "Shifting Your Brain". Online video clip. YouTube, 4 Sept. 2012: https://www.youtube.com/watch?v=5eqAXD_Loic

Sandel, Michael (2007) *The Case Against Perfection. Ethics in the Age of Genetic Engineering*, Cambridge, MA: Harvard University Press.

Savulescu, Julian (2001) "Procreative Beneficence: why we should select the best children", *Bioethics* 15(5/6): 413–426.

Savulescu, Julian, Bennett Foddy, and Megan Clayton (2004) "Why We Should Allow Performance Enhancing Drugs in Sport", *British Journal of Sports Medicine* 38: 666–670.

Savulescu, Julian (2005) "New Breeds of Humans: the Moral Obligation to Enhance", *Reproductive BioMedicine Online* 10, Supplement 1: 36–39.

Scheutz, Matthias (2012) "The Inherent Dangers of Unidirectional Emotional Bonds between Humans and Social Robots", in *Robot Ethics. The Ethical and Social Implications of Robotics*, eds. Patrick Lin, Keith Abney, and George A. Bekey, Cambridge, MA: MIT Press, 205–221.

Scottish Council on Human Bioethics (2006) 'Embryonic, Fetal and Post-natal Animal-Human Mixtures: An Ethical Discussion', *Human Reproduction and Genetic Ethics* 12/2: 35–60.

Searle, John (2010) *Making the Social World*, Oxford: Oxford University Press.

Shapiro, Gary (1980) "The Rhetoric of Nietzsche's Zarathustra", *Boundary* 2/8: 165–189.

Sharon, Tamar (2014) *Human Nature in an Age of Biotechnology*, New York: Springer.

Singer, Edgar Arthur (1924) *Mind as Behavior and Studies in Empirical Idealism*, Columbus, OH: R.G. Adams and Co.

Smith, John Maynard (1998) "Science and Myth", in: *The Philosophy of Biology*, eds. David L. Hull and Michael Ruse, Oxford: Oxford University Press, 374–382.

Sorgner, Stefan (2009) "Nietzsche, the Overhuman, and Transhumanism", *Journal of Evolution and Technology* 20/1: 29–42.

Stevenson, Charles Leslie (1938) "Persuasive Definitions", *Mind* 47: 331–350.

Stock, Gregory (2003) *Redesigning Humans*, London: Profile Books.

Swedlow, Jill (1999) *Great Dane. Model of Nobility*, New York: Howell Book House.

Symons, Michael (2000) *A History of Cooks and Cooking*, Urbana/Chicago.

Tännsjö, Torbjörn (2009) "Medical Enhancement and the Ethos of Elite Sport", in *Human Enhancement*, eds. Julian Savulescu and Nick Bostrom, Oxford: Oxford University Press, 315–326.

Taylor, James Stacey (2009) *Practical Autonomy and Bioethics*, New York/London: Routledge.

The Jesus Alien (2009) "Michael Jackson is not Dead", WordPress, 12 August 2009: http://www.thejesusalien.com/michael-jackson-is-not-dead-hes-an-ancient-egyptian-princess/

Thompson, Paul B. (2008) "The Opposite of Human Enhancement: Nanotechnology and the Blind Chicken Problem", *Nanoethics* 2: 305–316.

Tirosh-Samuelson, Hava (2012) "Transhumanism as a Secularist Faith", *Zygon* 47/4: 710–734.

Tooley, Michael (1972) "Abortion and Infanticide", *Philosophy and Public Affairs* 2/1: 37–65.

Treanor, Paul (2015) "Bimboid utopia: what do men want?", https://politicalaspects.wordpress.com/2015/02/06/bimboid-utopia-what-do-men-want/

Trotsky, Leon (2005) *Literature and Revolution*, Chicago.

Turkle, Sherry (2011) *Alone Together. Why We Expect More from Technology and Less from Each Other*, New York: Basic Books.

Twine, Richard (2013) "Animals on Drugs: Understanding the Role of Pharmaceutical Companies in the Animal-Industrial Complex", *Journal of Bioethical Inquiry* 10/4: 505–514.

Val, Jaime de, and Stephan Sorgner (2010) "Metahumanist Manifesto", www.metahumanism.eu.

Van Inwagen, Peter (2001) "Genes, Statistics, and Desert", in *Genes and Criminal Behavior*, eds. David Wasserman and Robert Wachbroit, Cambridge: Cambridge University Press.

Veenhoven, Ruut, and Yowon Choi (2012) "Does Intelligence Boost Happiness?", *International Journal of Happiness and Development* 1/1: 5–27.

Vinge, Vernor (1993) "The Coming Technological Singularity", https://www-rohan.sdsu.edu/faculty/vinge/misc/singularity.html.

Walker, M.A. (2009) "The Case for Happy-People Pills", *Free Inquiry* 29/5: 33–36.

Wellcome Collection (2012) *Superhuman. Exploring Human Enhancement from 600 BCE to 2050*, London.

Wells, Herbert George (1921) *The Island of Dr. Moreau*, London: Heinemann.

Wells, Herbert George (1967) *A Modern Utopia*, Lincoln: University of Nebraska Press.

Whitby, Blay (2012) "Do You Want a Robot Lover? The Ethics of Caring Technologies", in *Robot Ethics. The Ethical and Social Implications of Robotics*, eds. Patrick Lin, Keith Abney, and George A. Bekey, Cambridge, MA: MIT Press, 234–248.

Wills, David (2008) *Dorsality. Thinking Back through Technology and Politics*, Minneapolis: University of Minnesota Press.

Wilson, Edward O. (1978) *On Human Nature*, Cambridge, MA/London: Harvard University Press.

Wittgenstein, Ludwig (1953) *Philosophical Investigations*, Oxford: Basil Blackwell.

Wolfe, Cary (2010) *What Is Posthumanism?*, Minneapolis: University of Minnesota Press.

Wood, Gaby (2002) *Living Dolls. A Magical History of the Quest for Mechanical Life*, London: Faber & Faber.

Wright, W.D. (1998) *Racism Matters*, Westport, Connecticut/London: Praeger.

Wudarczyk, Olga, Brian D. Earp, Adam Guastella, and Julian Savulescu (2013) "Could intranasal oxytocin be used to enhance relationships? Research imperatives, clinical policy, and ethical considerations", *Current Opinion in Psychiatry* 26/5: 474–484.

Zak, Paul J. (2012) *The Moral Molecule. The new science of what makes us good or evil*, London: Transworld Publishers.

Index

M. Hauskeller, *Mythologies of Transhumanism*,
DOI 10.1007/978-3-319-39741-2

FSC
www.fsc.org
MIX
Papier | Fördert
gute Waldnutzung
FSC® C083411

Kolibri 360 GmbH
Ferdinand-Jühlke-Straße 7
99095 Erfurt, Deutschland
produktsicherheit@kolibri360.de